POWERS OF TEN

POWERS OF TEN

A book about the relative size of things in the universe
and the effect of adding another zero

Philip and Phylis Morrison
and
The Office of Charles and Ray Eames

based on the film *Powers of Ten*
by The Office of Charles and Ray Eames

SCIENTIFIC AMERICAN LIBRARY

A division of HPHLP
New York

Library of Congress Cataloging in Publication Data

Morrison, Philip.
Powers of ten.

Bibliography: p.
Includes index.
1. Cosmology—Miscellanea. 2. Powers of ten
(Motion picture) I. Morrison, Phylis. II. Office of
Charles and Ray Eames. III. Title.
QB981.M865 1982 500 82-5504
ISBN 0-7167-6008-8

Printed in the United States of America

Scientific American Library
A division of HPHLP

Distributed by W. H. Freeman and Company,
41 Madison Avenue, New York, New York 10010,
and 20 Beaumont Street, Oxford OX1 2NQ, England.

Second printing 1996, HC

To Charles

R. E.

The concept, the form, the most striking pictures in this book are part of the luminous legacy of a man who never saw it. We dedicate the book to that man, whose clarity and delight informed so many visions.

To Charles Eames, Architect

P. M. & P. M.

The idea of exponential series was a favorite tool of Charles Eames, as an architect who loved the art of estimation; he saw it as a way of bringing very large quantities within our grasp. In the 1952 film *A Communications Primer*, the use of powers to help visualize large numbers was introduced; and one of the early mathematical peepshows, made in 1961, was about the powers of two.

Charles liked to quote Eliel Saarinen on the importance of always looking for the next larger thing—and the next smaller. The idea of scale—of what is appropriate at different scales, and the relationships of each to each—is very important to architects. Kees Boeke's book suggested the possibility of bringing these ideas together in film. With a constant time unit for each power of ten, an unchanging center point, and a steady photographic move, we could *show* "the effect of adding another zero" to any number.

The Morrisons, Phil and Phylis, had the imagination and enthusiasm to see what this idea could mean. During the making of *Powers of Ten*, ideas poured out, fascinating information piled up—more than could ever be fitted into the constraints of the film. But now, in book form, the opportunity exists not only for the straight line journey, but also for enriching additions and insights at each power, for all our pleasure—expanding our understanding, expanding our knowledge. And in book form it is possible to start at the human scale of one meter, moving outward, power by power, and understanding little by little what it means to add another zero—or to start at the farthest mysterious point in the journey, moving through the human scale of one meter down to the smallest scale yet explored.

RAY EAMES

Powers of Ten is a phrase you will hear soon enough in almost any scientific conversation. It is also the short title of a brief and beautiful film produced by the Office of Charles and Ray Eames. We came to know the Eameses and their studio through taking part in the filmmaking, first at some remove but later quite intimately. You can sample the film in this book, as much as can be expected of a stationary representation of what is meant to stream by at twenty-four frames per second.

This book is a transformation of the film, as the film was itself a transform of an earlier little book, *Cosmic View: The Universe in Forty Jumps*, by a Dutch school teacher. Kees Boeke's innovative book for children was our introduction to this ingenious itinerary, and we treasured it for years. That journey—in the book a series of jumps, in the film a disciplined smooth flow—became for us a journey throughout science.

These pages present a like voyage of discovery through the universe, but rather more freely. The original trip was a long and uninterrupted straight line: the changing view along the line was presented without a sidewise glance. Instead, with Ray we have organized a set of illustrated comments, lingering here and there, looking around, recalling the experience of earlier travelers, and seeking to convey the evidence that informed the carefully constructed images of the journey itself. The unity that arises so clearly out of the diverse scenes will we hope become perceptible and exciting to readers new to science, young and old alike. We have tried particularly to make easily understandable the world we know from instruments and inference alone. At the same time, we are confident that the well-grounded reader will find it as satisfying as we ourselves did to see so much laid out in context scale by scale. Anyone can see how much we draw on the community of science; we have more compiled a guidebook than written an original account.

Finally, this book belongs among the first written of a new series of books intended to expand upon science as it has been displayed in worldwide growth in *Scientific American*. We hope that this small book and its companions will be as warmly and widely used as the magazine itself has been during the years in which we have helped prepare—and then always excitedly read—its monthly pages.

PHILIP AND PHYLIS MORRISON

CONTENTS

POWERS OF TEN

To describe the evolutions in the dance of these gods, their juxtapositions and their advances, to tell which came into line and which in opposition, to describe all this without visual models would be labor spent in vain.

—Plato, *Timaeus*

LOOKING AT THE WORLD: AN ESSAY

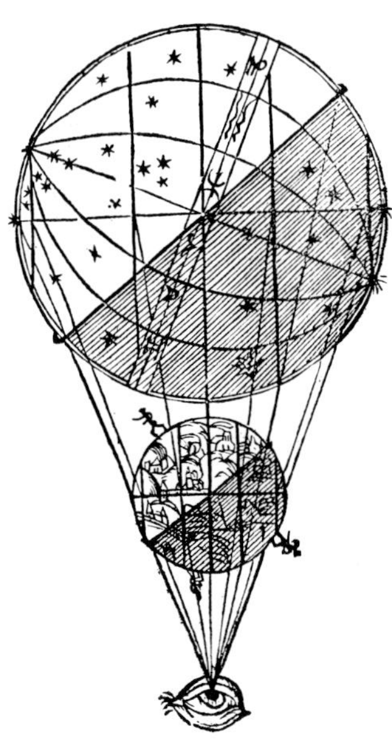

Of all our senses it is vision that most informs the mind. We are versatile diurnal primates with a big visual cortex; we use sunlit color in constant examination of the bright world, though we also can watch by night. Our nocturnal primate cousins mostly remain high in the trees of the forest, patiently hunting insects in the darkness.

It is no great wonder that the instruments of science also favor vision; but they extend it far into new domains of scale, of intensity, and of color. Inaudible and invisible man-made signals now fill every ordinary living room, easily revealed in all their artifice to ear and eye by that not-so-simple instrument, the radio, and the even more complicated TV set. It is very much this path of novelty that science has followed into sensory domains beyond any direct biological perception. There, complex instruments assemble partial images of the three-dimensional space in which we dwell, images rich and detailed although at scales outside the physical limits of visible light.

The images finely perceived by eye and brain in a sense span the scientific knowledge of our times (though it is risky to neglect the hand). The world is displayed by our science in diverse ways, by manifold instruments and by elaborate theories that no single person can claim any longer to master in all detail. The presentation of the whole world we know as though it were a real scene before the eyes remains an attractive goal. It should be evident that no such assemblage could be complete, no picture could be final, nor could any image plumb the depths of what we have come to surmise or to understand. Behind every representation stands much more than can be imaged, including concepts of a subtle and often perplexing kind. Yet it is probably true—truer than the specialists might be willing to admit—that the linked conceptual structures of science are not more central to an overall understanding than the visual models we can prepare.

The Enlightenment used the principle of projection in a matter-of-fact map. The familiar stars of the sky were made to enclose the newly detailed earth, well mapped without benefit of orbit.

THE GAMUT OF THE SCIENCES

The world at arm's length—roughly one meter in scale—is the world of most artifacts and of the most familiar of living forms. No single building crosses the kilometer scale; no massive architecture, from pyramid to Pentagon, is so large. A similar limit applies to living forms: The giant trees hardly reach a hundred meters in height, and no animals are or have ever been that large. The smallest individual artifacts we can use and directly appreciate—the elegant letters in some fine manuscript, or the polished eye of a fine needle—perhaps go down to a few tenths of a millimeter. Six orders of magnitude cover the domain of familiarity. Science conducted at these scales is rather implicit: The most salient disciplines are those that address the roots of human behavior.

Let us begin to marshal the furniture of our world according to its physical size. On larger scales, only occasionally does the work of our energetic species show up: a bridge, a wall, a dam, or a highway. These are typically less than fully three-dimensional. They seem long ribbons when occasionally they are caught in aerial views. Only in their collectivity do we see human artifacts that occupy large surface areas (still not three-dimensional) in the ten- to hundred-kilometer range, sometimes even beyond. These are the cultivated plains and terraces, the irrigated lands, the clearings of the ancient forest, the great cities and their environs. Theirs is a history of growth more than one of design. For the rest of life, too, we find a similar display. Blades of grass are small, but grasslands and savannahs, like the dark forests north and south, extend over whole regions, easily up to a thousand kilometers across. It is these regions that make up the visible large-scale landscape. Here the cognizant sciences are those that aim at the nature and use of lands. Perhaps still more germane are the descriptions offered by the historians and geographers of an earlier time, and those offered by the adepts of elaborated practical technologies, from agriculture to forestry and engineering, ancient and modern alike.

Once past the scale of a thousand kilometers, we lose sight of our species. At the global and regional scale, up to ten thousand kilometers, cooler sciences enter in strength. The swift motions of the air, its clouds and ceaseless winds, the slower flow of rivers, ocean currents, glacial ice, and the majestically slow drift of the solid continents themselves lie behind the single views. These occupy the dynamical sciences of meteorology, oceanography, hydrology, and geology. Within this generation, geology has far extended its grasp; until recently, the globe as a whole was hardly a geological topic. Regions were well understood, but no known single process had joined distant shores of wide oceans or ringed the whole globe. All that has changed: Today a geologist may take the earth for province.

One leaves the earth, but not yet the domain of humanity, beyond the range of ten thousand kilometers. Out to the moon we have sent intrepid explorers, while the geosynchronous earth orbit, a ring five earth radii out into space above the equator, is now a well-exploited natural resource. Satellites orbiting within that gravitational band neither rise nor set, as watched from the spinning earth, but remain always in view of the artfully aimed fixed dishes; they relay word and image over radio links to and from almost every nation.

It takes a scale six powers of ten larger to reach the boundaries of our solar system, out there among the unseen comets. The sciences of the solar system—the studies of the surfaces and the interiors of the planets, large and small, their satellites, meteorites, the comets, the dispersed dust—are today more than just astronomy. No longer do we merely look from afar; we touch and sample now, at least vicariously with our robot probes. Astronomy proper begins now with the stars; one of them, the sun, is our own life-giving hearth, the only star close enough to permit detailed study. A great gulf is open between our home region near the sun and even the second closest star; the steps must cross four or five powers of ten to enter the realm of the stars. It is a remarkable story, first told in our century: the birth, development, and life history of the stars, diverse globes of gas into which most of the visible matter in all the universe is bound. This is the kind of astronomy held in the root of that word itself: the study of the stars. It is mature now, though certainly unfinished.

Let us now look the other way, inward from the submillimeter world of the attentive but unaided eye to the microcosm. First in interest come the intricate machinery of our own bodies and its counterparts in all the larger forms of life. Here we engage anatomy, physiology, histology, cytology—a battery of specialties, ending with the study of the cell itself, the ubiquitous unit of living forms. Three or four further powers of ten span that whole microscopic world of life—microbiology—down to the smaller cells of the most ancient forms of life, to uncover the not quite living parasites, the viruses. But at that level, on the scale of a thousand angstroms or so, we encounter the mechanisms of molecular biology (and its newer emulation in the textured artifacts of microelectronics). These scenes relate form to function: The form is molecular; the functions are among the deepest properties of life, shared by the full web of life during all the time we now know of terrestrial evolution. Here we speak of genetics and of the biochemistry of large molecules and their cycles of interaction. Before long we cross the vague frontier that parts life itself (so much the most subtle of chemical processes) from the chemist's world of random motion and atomic bonds.

Look out again at the celestial scale. There too we cross a real boundary of nature once we leave the Milky Way environs to see galaxies, whole spinning pools of stars bound together over time. The astronomy of the stars first extended to the dilute interstellar medium, the star-forming matter from which new stars are born, and then went on to galactic and extragalactic astronomy. The fascinating forms and varieties of star-pools are strewn throughout space as far as we can see into the distant world.

Again in journey inward from the world of the large molecules, we reach at last the individual atom, at a scale of about an angstrom. Below that scale, all science is physics and chemistry; once we set out to explore the innermost spaces of the atom, we enter a strange domain well beyond any direct imagery. We can represent it only as it has been puzzled out with the powerful tools and concepts of modern physics. Our study has led far, to new fundamental laws, at first paradoxical but now fruitful in explaining the patterned and stable world we know within matter. The modular world of the hundred chemical elements and their larger but still limited variety of nuclear species is a world ruled by a subtle interplay between order and chance.

The two ends of our procession of images, the terminal scales of large size and small, mark the limits only of contemporary knowledge. On the one end, far out where the galaxies appear like a glowing froth in darkness, all our sciences become only one: cosmology. We know of no spatial novelties beyond the billion-light-year point. All the distinct structures we are aware of are safely smaller than that. There is certainly wonderful novelty, but it is expressed not spatially but over time.

The universe has come to be filled with those diverse galaxies formed out of a once bland and uniform stuff. At the other end, for the very small we have again one science only: particle physics. There are even hints that the two ends inform each other: The fiercely hot early universe may once have held only the kind of matter we see transiently now in the particle labs. Ours is a modular world, built out of myriad replications of the simplest structures, structures that we are only now beginning to understand. Within the nucleus is the proton; within the proton, the interacting quarks. Within quarks? The magnetized rings and tubes that are our great accelerators, the ultramicroscope probes of our time, have not given the final answer.

Forty-two powers of ten so far span our firm knowledge; we have only brave hints and conjectures beyond that. We do not yet know, though we can argue about it, whether infinity lies within the real world as it lies within the mind's reach. How far can we continue the excursion in either direction, out to the macrocosm or in to the microcosm? Some day we hope to learn.

THE SAME AND THE DIFFERENT

Over the wide sweep of size, the eye scans all that is made. When the thought of Aristotle turned to such a gamut, he saw the sky-world as wholly different from this earth on which we dwell a little while. We know his distinctions: up there, the objects are shining, circular, perpetual. Here below the moon, matter is usually dark, and its motion is neither elegant nor enduring. But in our times we have seen these two worlds made one. The satellite we launch is

"L'encyclie du monde," presented in the sixteenth century as a complex of the five elements, encompassed by the heavens.

earthy, faulted, marked with the finiteness of the minds that designed it and the hands that wrought it. Yet given that one terrible flaming impulse of the rocket, it enters the celestial regime, into an orbit that can be as purely circular as that of any celestial body, there to shine enduringly, bright as a star in the night sky.

The world is not dual; that below and that above are one and the same. The difference is but distance and motion. Aristotle's theory, which has its insights, was fully superseded by the wider insights of Copernicus three centuries ago. Since then we have come to know well that our earth is just as celestial as any other planet; it too shines, its motion is circular and perpetual—if epicyclic—once viewed from far enough away, once seen from elsewhere in the sun's cortege. We dwell on Spaceship Earth. By the same set of experiences, we recognize that the red wandering star we call Mars is not an ineffably luminous body, quite unlike the earth we know. Our cameras show instead that it looks a good deal like the desert of Arizona. All the rest are places, too, places as physical, as finite, as wonderfully complex as earth. Of course they are not identical to earth; our sun, for example, is a giant fiery ball, shining and hot as no planet can be. But its globe of gas gravitates just as do the rocks of earth; its nuclear fires, so enduring as to have warmed our earth since its earliest days and to have nourished somehow out of water and air the thread of life, are now under detailed study. The solar fires are long-lasting but finite; one day we may set up their counterparts here (as we have already done transiently in grim thermonuclear explosions) to run the shafts of industry. Heaven and earth are not wholly distinct, nor are they one; they differ in their nature, yet they are joined in a wider unity.

So it is with the microcosm. Again the Greek insight was wise. The early philosophers tried to explain how wood could change to fire and ashes, how bread could nourish the hungry, how black iron might rust red. They had the idea that deep down below the size we perceive, matter was a web of small modules—the atoms—whose incessant rearrangements account for all becomings. That profound idea, turned from speculation to a powerful and growing mastery of the hundred elements, has teased out the linked fabric of every substance old and new. It has led to today's view of atomic matter, made clear in the images we build of the world of the small. Again, there is no trivial path to that understanding: The atomic world is not just like the one in which our senses place us. To be sure, it is the same world, for we have found no other anywhere, but it is related to the world of familiar experience through that same curious blend of the marvelous and the homely that we find out there among the planets. The step-by-step examination within these pages is best shared by a traveler who is pleased alike by unexpected familiarity and by exotic novelty.

LILLIPUT AND BROBDINGNAG

Dr. Lemuel Gulliver, acting for the ironic Jonathan Swift, is the most famous of travelers among the powers of ten. (Truth to tell, he visited the two neighboring powers of twelve.) The pettiness of the sectarian wars in Lilliput and the disdain of the magnanimous Brobdingnagian giants for us self-serving little humans are both deftly reported as a new quality of moral judgment arising from size alone. In physical dimensions, these peoples, big and little, differ from us mainly through our perceptions of them. Gulliver sees the Lilliputian world at a distance, say to admire the elegant threading of an invisible needle with invisible thread. The giants he sees as if in a microscope; their most elegant artifacts seem crude, their very bodies gross and

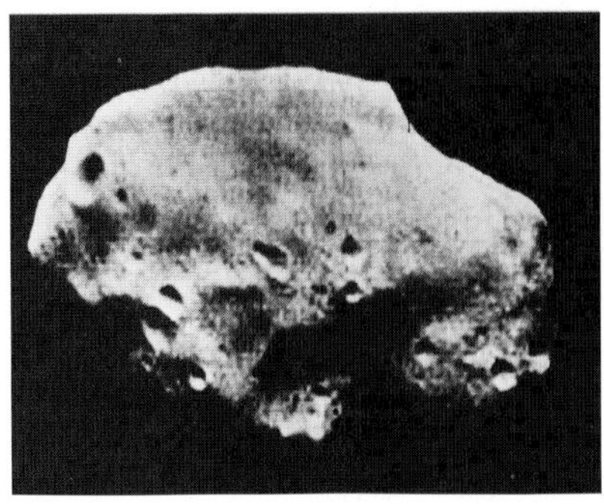

The gravitation of Phobos, a tiny moon of Mars only about ten kilometers across, is too weak to draw the rocky structure into a sphere.

marred to his too-close gaze. Exactly this had been the real surprise offered by the microscope: The finest needle points turned jagged and coarse in its magnifying lens, and the most admirable complexion viewed microscopically was a sea of blemishes.

But the effects of scale go well beyond perception in engendering novelty. The world works differently at different scales. Swift well knew the foundations of geometry; he understood that it takes not simply twelve, not even twelve times twelve, but twelve times twelve times twelve six-inch Lilliputians to equal the bulk of one six-foot man. For not only must the height be increased twelvefold but the width and the length must be increased as well. (Eight cubes of sugar—two layers high, each layer a two-by-two square—need to be stacked together to build up a single cube with an edge twice as long as the edge of one of the original cubes.) Thus the Lilliputian planners requisitioned for their prisoner Gulliver 12 x 12 x 12, or 1728, daily rations.

Experience shows that so simple a trust in schoolbook geometry is unrealistic. Consider that a man might get by on a loaf or two of bread a day. But to feed a small animal like a chipmunk, six inches long, on about one part in two thousand of that diet means the active little creature is fed each day a piece of bread no larger than your thumbnail. Starvation rations! The animal will contentedly eat a third to a half of a slice each day.

For the truth is that the world is not ruled wholly by the simplicities of Euclid. A scale model may resemble its counterpart with fidelity to the eye, but, in general, the model cannot work in the same way. It was Galileo himself who first pointed out this result. Put aside the complexities of nutrition, heat loss, and the rest, and consider one single static but important property—structural strength. The learned Salviati asks early on the first day of the dialogue in Galileo's graceful *Discourses Concerning Two New Sciences* (1638), "*For who does not see that a horse falling from a height of six or eight feet will break its bones, while a dog falling from the same height . . . will suffer no harm?*" Later he remarks, "*A little dog might carry on its back two or three dogs of the same size, whereas I doubt if a horse could carry even one horse of the same size.*" Indeed, small things are relatively strong, large ones are weak. Great steel bridges cannot support a load equal to their own weight; any wooden plank does much better. It is no special miracle of design that fits a worker ant to drag back to her nest a fly much bigger than she is, nor one that allows a small bird to fly so well, while humans can barely fly on muscle power but must enlist the aid of hot and thirsty metal engines. These facts imply that form follows not function alone but size, especially over large changes of scale. Such is the effect, in the physical world, of "adding another zero."

We cannot follow up the reasons for all the effects of scale change; that would require treading most of the paths of theoretical science. It is enough to see how one case works out: the structural weakness of the great and the strength of the small. Every structure on earth must support itself against the pull of gravity. Gravity is exigent; it reaches inside every candy box, no matter what the wrapping, to distinguish the full pound from the empty container. Humans, horses, and dogs walk about on their feet; only the foot surface is available to bear the full weight of the whole organism. That surface area, like all areas, increases by 10 x 10—a hundredfold—for one tenfold step in linear size, say from a small dog to a horse. So the bearing area goes up by one hundred. But the weight of the

Dione is a moon of Saturn, seen dramatically lit against the big planet; five hundred kilometers across, it is exquisitely spherical.

whole body, the total load to be held by the bearing surface, goes up by 10 x 10 x 10—a thousandfold. The horse is sure to have proportionately less structural reserve than a dog. Whatever its design, a structure enlarged sufficiently must therefore fail. Where the design remains similar in form, the reckoning is accurate. Of course a dog is not in fact built very like a horse. That difference in form partly reflects the distinct differences in behavior of dog and horse, but it partly reflects the adaptation to a change in scale. Large animals tend to thick and sturdy form; small animals are recognizably graceful, agile, subject to chill, restlessly hungry, easily waterlogged. Each of these familiar traits can be given at least rough explanation as a simple demand of scale.

There is an inner reason for the difference. Dog and horse are built of the same materials, of flesh and bone. Deep inside matter, the atoms do not change in size as the size of the creature increases. The scale model is inconsistent. The stamp of intrinsic size is held in the nature of the atom; in another universe than ours, one with differing atoms, these arguments might be circumvented. But here they rule. Nor is it only the elegant engineering of living forms that reflects the decisive result of adding zeros. Human technology naturally must obey the same rules. Moreover, natural forms, quite apart from life and its evolution, are no different; mountains too must bear their own weight. They can reach heights only very small compared to the whole earth, or even to Mars. Every planet, every star is nearly a sphere. But smaller bodies, like the little moon of Mars called Phobos, can be quite unlike a sphere: Phobos resembles a rather poorly formed potato. The reason is in its size alone. As the size of a body increases (usual materials always assumed), the self-attraction arising from gravitational forces pulls internally on all portions of the object. Let the body become of great size, and no material has sufficient strength to withstand the effects of gravity. The object is self-compressed; it strains to become as compact as possible. Once it is massive enough, it must approximate a sphere. It turns out that the yield limit for rock is reached somewhere near a couple of hundred kilometers; below that size, satellites and asteroids come in all shapes: bars, bricks, lumps. Much above that limit, they are all spherical. Astronomy is thus the regime of the sphere; no such thing as a teacup the diameter of Jupiter is possible in our world.

Let us look at still smaller objects—and at other properties. No one would expect that, if you dipped your hand into a bowl of sugar cubes at the table, you would bring out a cube with it, unless you positioned your fingers to grasp the cube. But if the bowl instead contained granulated sugar, some crystals would surely stick. If the bowl held powdery confectioner's sugar, you would fully expect to withdraw a white-dusted sugary hand from the container. Why? It is all the same material; the adhesion of sugar to skin remains the same. But that adhesion grows with the area of contact between the two surfaces. The total stickiness increases with the area of the sugar sample removed; but the weight of the sugar varies rather with its volume. Normal adhesion cannot lift a sugar cube: too much volume, too little surface. For the powder, the circumstances are reversed. The world under the microscope is dominated by just such surface effects; it is frictional, sticky, clinging. Small bodies do not coast; they show little effect of inertia. Bacteria experience pure water as we would a pool of honey; the water is the same, but the cell's surface is so large by comparison with its tiny volume that even the water we find so yielding fiercely inhibits their motions. We repeatedly encounter phenomena associated

A journey through space alone misses something important: change. Here is the picnic scene as it was seen in an aerial view in 1933, when the World's Fair was in preparation.

with scale change throughout our journey. Stars shine, planets are round, bridges remain geologically rather small, cells divide rapidly, atoms randomly vibrate, electrons disobey Newton, all because of scale.

INVISIBLE ORDER

No visual model can convey unaided the full content of our scientific understanding, the less if it is restricted to the static. Pictures in a book mostly present a static account of the world, a limitation not imposed upon swift-flying film or videotape because of their vivid fidelity to the world of change. Film and the video processes together constitute the most characteristic form of art in this changeful period of human history. The limitation of the static image is not simply that it lacks the flow that marks our visual perception of motion: Real change in the universe is often too slow or too fast for any responses of the visual system. The deeper lack is one of content. A single take belies the manifold event. Several still images in a sequence chosen to bring out the nature of the change can often do as well cognitively, if not evocatively, as the flow of re-created motion itself. But that sequence is the key: It takes more knowledge to convey change. While our attention is seized by the moving image, we usually forget that it is nothing but a subtle illusion induced in eye and brain by a multiplicity of still pictures. But what a profligate flow it is! A couple of dozen distinct pictures each second are needed to simulate ordinary moving reality.

All that implies how much a visual model needs some additional means by which to present time, to mark change and its rates. For the world is a pageant of events whose relative rates of unfolding in time are as important as the size of the stage upon which they play.

One familiar kind of unfolding can offer a transparent example. The physical world is in motion, whether openly (as in our everyday surroundings) or more hidden (as it is within molecular matter). Sometimes the tempo is so slow that change is concealed from one quick glance, as in the larger cosmos. We measure the speed of motion most directly by noting by how much an object's position in space changes in a given time. The apparent speed of the planets in the sky provides a correct clue about distance that not even the most archaic of naked-eye observers would ignore. The perceived sequence of decreasingly swift motions is familiar: birds, clouds, moon, stars. . . . Since Copernicus, we have been able to translate the observed round of the planets—how many years each requires to

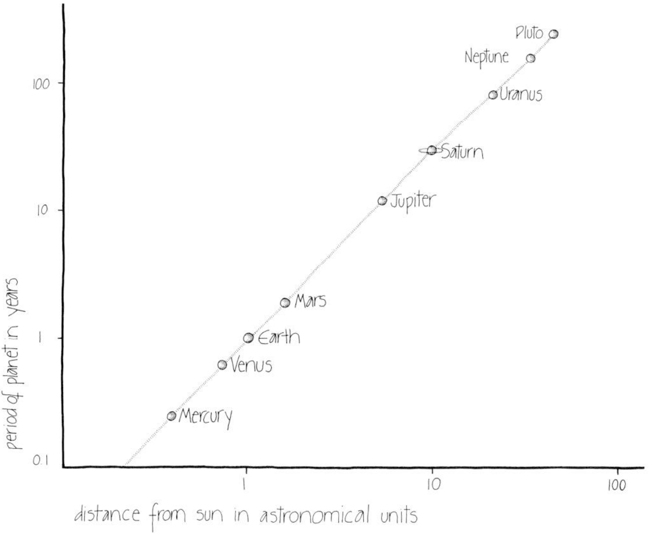

The surprising simplicity of the relation between the size of the planetary orbits and the time it takes the planets to traverse them is part of the invisible order. Noticed by Kepler, it determined the force law for Newton's universal gravitation.

return to the same point, taken not with respect to the earth, but to the central sun—into their true orbital speeds. For the size of their circuits can be read from a Copernican map, relative not to the meter standard but to the orbit of the earth. Jupiter's near-circle is five times the diameter of Earth's. But Jupiter's year is about ten times longer than our year. That result (even then known with accuracy) was in the hands of the Copernicans from the first; it falls at once out of a scaled diagram of nested circles. Copernicus's system is by no means as simple as his drawing might suggest. Copernicus found it necessary, as today's nonastronomical reader does not, to take into account the small deviations of the planets from uniform circular motion, long known to the old star-clerks.

The speeds turned out to be of supreme interest. In the figure, the measured relation of speed to orbit size is plotted in graphic form using the data of today (taking a mean radius to represent the actual ellipse). The regularity from planet to planet is striking; it is a relation first adduced by Kepler and made rational in detail a lifetime later by Isaac Newton. Our point is simply made: Without watching over time, we could not have found this rule. Nor is this merely one detail of order missed by the static onlooker; it is a profoundly diagnostic one. Compound this regularity among planets with a similar rule for the little moons of Jupiter, for those of Saturn, and for our moon itself (its motion related to the fall of an apple at the earth's surface) and something of the order behind celestial appearances comes through. Go on to artificial satellites, whose motion was predicted by Newton in the seventeenth century, and on to the distant double stars, and the point is universally made. Once time is added to space, once the visibly moving model is augmented with quantitative inferences, the invisible order of the world earns the awe with which the Enlightenment saw it. Their sense is shared by the physicist today, even one who recognizes fully that the world holds essential disorder, too.

The simplicity of the celestial rounds has long made them the paradigm of order in science. But the study of change has gone well beyond the planets. If the eighteenth century admired the Newtonian revolution to excess, our present thought is equally caught up in the physicist's notion of energy. Energy does not mean something close to sunshine, gasoline, or jelly doughnuts; the technical meaning is clear, but altogether more abstract. The concept has been extended successfully beyond the mechanical domain, in which it first arose, to every process of physical change. Energy is a quantity that can be defined for any system; its value can be calculated, but not directly observed, by procedures that demand much detailed analysis of the system. Once the value is found, it holds without change during all the changes any system can undergo: thermal, chemical, biological. . . . Only if the system is in some sense open (that is, if somehow energy can flow along specific paths out into or back from the world outside) can the energy content change. Even for such an open system, it is always possible to seek out the compensating change in the environment that, once included, leaves the total energy content of system and its surroundings again invariant. The aphorism of the nineteenth-century physicist Rudolf Clausius, a pioneer in the wider understanding of the concept, holds today: "*The energy of the world is constant.*"

To energy one should add a variety of other examples of constancy, like momentum along a line, momentum of rotation, and electric charge, which so far seem to hold unreservedly. Their utility is wider than that of Newton's laws or of any other of the more detailed statements of chemist or physicist: They turn out to be as true and indeed as useful in the quantum theory of the nucleus as they are in the study of the distant galaxies. For many complicated natural systems, they imply the major quantitative statements we can make. Despite any entrepreneurs who may now be busy selling perpetual motion devices, no physicist fears the failure of these high principles. To be sure, for the universe as a whole—if that somewhat paradoxical concept can be as well defined as cosmologists now think—limits may someday appear even to this class of generalization. Short of that, the invariants work always and everywhere, at every one of the forty-plus orders of magnitude. These invariants (there are more) find a rational connection with plausible properties of space and time; they imply a consistency we tend now to demand of proposed new theories of any kind. Einstein and Schrödinger offered their novel theories, far from Newton's paradigm, but the conservation laws hold in those new theories too in quite recognizable ways. Yet they cannot appear in the model held before the eye, save by implication.

Such invisible regularities can reveal the invisible, even literally. A few star images seem to undulate along the line of motion very slightly as the years go by, following a path like a gently meandering stream. Why? The momentum law explains that such a mass cannot behave so by itself; there must be an unseen companion to share the momentum, now pulling the visible star to one side, then to the other. Disturbances were found in the motion of the planet Uranus in the 1840s; on that evidence, an unseen planet was predicted by John Couch Adams and Urbain LeVerrier, promptly found in the telescope, and named Neptune. The last major planet to be discovered, Pluto, was forecast in a similar way. (Pluto turns out to be so small that we now believe the original data were inadequate, and the final success of the search was a tribute more to patience than to Newton.) What worked for planets worked for atoms too. In 1932, certain cloud-chamber tracks showed protons recoiling —from nothing. No initiating track was seen toward the proton from the novel radioactive source outside. But the inference was clear, exact, and soon verified: James Chadwick disclosed the neutron by directly applying the same notion that had found Neptune. It is this catholicity of scale—planet to nucleus—that warrants the conservation laws as the strongest pillars of the invisible order.

Color is part of the visual experience, like change. But the perception of color is no simple accomplishment of the eye and brain. Rather, it usually includes an unconscious internal comparison over the whole visual field, to allow for the quality of the light that illumines the scene. The basis of color, the physical quality that distinguishes red light from blue, is the energy carried by light in each single interaction with atomic matter, ultimately in the retina. All those forms of energy that are strict analogues to light—from radio waves to gamma rays—can also be ordered by the interaction energy they carry. The variations in energy are large—sixteen or eighteen powers of ten—from the very low energies in radio to the highest-energy gamma rays. (The visible gamut from red to blue spans only a two-fold increase in energy.) These rays all move at the speed of light in empty space; but they interact differently with atomic matter, according to their intrinsic energy. At the low-energy end, the interaction is diffuse, gentle. Energy carried along into matter by radio—though it can be great overall—requires a huge number of single interactions. At the gamma-ray end, the same total energy would express itself through only relatively few atomic interactions.

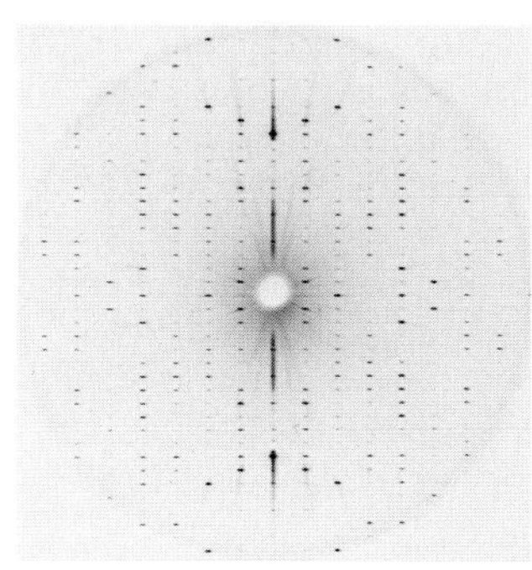

Another part of the invisible order appears in these spots, the photographic record of the orderly diffraction of x rays of a single energy from a crystalline sample of DNA. Measurements of such a spot array led to our knowledge of the form of the double helix. The technique amounts to a powerful but indirect x-ray microscope. The sample here is a newly found form of DNA, a curious left-handed double helix, whose function may bear on the control of development in many-celled organisms.

All these types of radiation are called electromagnetic because they interact solely with electric charge and its magnetic counterparts and with no other property of matter. The images they form are subject to strict geometric limitations. Low-energy radio is incapable of revealing fine structures; its limits are macroscopic. High-energy rays can probe deep within single material particles, but they disturb the receiving particles by their disruptive transfer of energy. The limit for spatial discrimination is proportional to the energy carried. This limit is roughly the wavelength of the radiation, to describe it in language that expresses the wavelike and not the raylike aspect of the radiant energy. Individual atoms of angstrom size can be imaged conveniently by radiation whose size discrimination is at their own scale; that is, the x rays. Quite elegant means have been found to recover the spatial arrangement of a crystal exposed to x rays from the record of the redirection of the radiation as it passes through the repetitive array of atoms. That is the technique that maps the forms of such molecules as DNA. The reconstruction is carried out by computation; it is simply impractical to devise lensing systems that work well enough for the x-ray region.

The fiery color play of the opal is like x-ray diffraction at another scale. An opal is a stacking of tiny balls of silica, each about the size of the minimum discriminating distance of visible light. Its effect on such light is to disperse it into different directions, which depend upon the energy of the interacting light. The different colors are thus separated, and the fire results. A careful record made with light of one sharp energy—one color—would produce, instead of the color play, a geometrical pattern of bright spots in that one color; that pattern could be analyzed to form an image of the opal array in space. The x rays produce the same phenomenon but on the scale of the atom itself.

It is extension of this same set of ideas to incoming beams of particles, electrons and even protons, that has allowed indirect study of the internal forms of the nucleus and of the subnuclear world. The method is there more complex because the results are so much more diverse. If light enters an opal, only light comes out. But if a fast proton enters a nucleus, many new particles may emerge. Yet the ordering principles that depend on the transfer of energy, momentum, charge, and spin still control the events.

THE FINE PRINT IN THE ALMANAC

The refinement of astronomical observations is not a new story. The long time during which the motions of the planets have been studied has allowed us to store up a great treasure of data. In a specific city, a solar eclipse can be predicted a century in advance with an error measured in seconds. That precision has led students of the motions of the solar system to a concern with fine points. The simple story of circular orbits in strict obedience to the unmoving sun is evidently too simple. It describes reality well, but with a broad brush. We can speak of a model of the real system. From Newton's day onward, it was the task of the theorists to rationalize more and more precise observations by more and more attention to detail: improving the model. The orbits are clearly not circles but ellipses. That itself goes a long way toward better agreement. But they are not truly ellipses either. The ellipse is the strictly correct form of an orbit on which only the massive sun acts. Yet the

The orbit of the earth, or of any planet in the solar system, is not an accurate ellipse. It is always close to an ellipse, but the orientation and form of the near ellipse slowly changes with time, so that the true path is close to an unending open rosette, drawn here much exaggerated. For Mercury, whose ellipse shifts fastest, the orbital turning point moves around a full circle only after a hundred thousand orbits.

planets in fact pull upon each other and even on the sun itself; the full study of the overall motions must take some account of the influences of each of the planets. The slighter the disturbing effect to be accounted for, the smaller the planet and the farther off the source of the disturbance that must be considered. Calculations can include a great many interacting bodies by drawing on the full computing power of these times. The orbits are in reality complicated, slowly evolving, endless rosettes, always near a shifting ellipse but beyond any simple description. It is this step-by-step complication of a first approach that is the chief current task of many a science, though few have the power celestial mechanics possesses. The model of the circles has been inordinately refined.

Why push so hard the improvement of a model? Initial understanding would follow mainly from the circles and the simple curve of the graph. Why the rest, which demands most of the expertise and carries the effort into technical domains beyond the nonspecialist? There is the challenge, like that of the mountaintops. More than that, detail is sometimes a necessity. No one can aim a spacecraft at Mars and expect it to arrive there if the aim has been based on an approximation of circular orbits and the assumption that the only gravity that matters is that of the sun. So the unknown future is a strong incentive to model improvement. Most of all, no one can know what is hidden in the deviations; there is always some hope of a grand discovery. Indeed, Neptune was hidden in the small deviations of the motions of Uranus; a major planet or two is no small prize. The astronomers of the late Victorian decades looked hard for additional planets near the sun, because their orbital calculations of Mercury's motion were plainly infected by a small error that would not go away. No innermost planet has yet been found that might be responsible for turning the Mercury rosette just a hair faster to fit the now-certain facts; instead, that discrepancy has been accounted for fully by the new and radical view of gravitation that Albert Einstein presented in 1916. It was the first fruit of his most original accomplishment, the theory of gravitation known as general relativity.

It is the occasional grand consequence of a minor misfit that has formed a good deal of the popular caricature of theoretical change in natural science: a succession of schemes of the universe, each one all-embracing, each to be shattered beyond repair by an ocean of change that floods through a single crack. No one would deny that this view contains some truth, but it is hyperbolic. We no longer need, like the philosopher Kant, to imagine that human thinking itself requires Euclid's geometry and Newton's laws of motion. Nor is the universe seen as a clock of infinite precision, a clock that, once wound up, is fated to tick out its inexorable rhythms. We are more modest today in physics and probably much closer to reality: There is room for strict cause as there is place for a reasoned uncertainty, noise, and chance. The two together play out the intricate universal drama. It is true that Einstein's theory lends itself well to difficult talk of curved spaces and the geometrization of space-time, and those insights are even irresistible to a degree. But, in our visual model of the world, such effects can hardly be seen, for the old descriptions of force and motion on a large scale (even if wanting when pressed too far) neatly account for most of what we see within the larger cosmos, out to a billion light-years and more.

Let it be affirmed that gravitation, in the plain old Newtonian style, is today by no means the effete and elegant subject of precisionists. Their studies saw no grandly new phenomena but only agreement in the last decimal places: fine print in the almanac. One can easily take that inference from the history of celestial mechanics: Nothing very new is to be expected. But that is not true. Gravitation is still full of surprises, though to be sure not in the motions of the major planets. Their orbits have mutually smoothed themselves out in space over the long repetitious span of solar time. But look at the galaxies in collision, their long arms and counterarms telling a strange yet purely gravitational tale. Even the new compact objects, like the neutron stars, are understandable as the outcome of Newtonian pull, no less than the intricate rings of the planets are. The wild comets, never disciplined to the flat plane of the solar system, are occasionally imprisoned for indefinite sentences inside Jupiter's orbit. All these are signs of the versatility of gravitation. In effect, the whole world of the large is one drama of resistance to the inexorable, never saturated, never forgiving pull of gravity, from which its orbit, its high temperature, or its quantum motion may preserve a system, at least for a while.

QUANTUM MOTION: SURPRISE AND COMMONPLACE

If fifteen of the largest powers of ten within our visual grasp are dominated by gravitation and its richness, several scenes at the small end are examples of quantum motion. They follow new laws, not those of Newton; the effects of gravity are all but absent. The domain of the atom—both in its external partnerships, up to the scale of molecular mechanisms, and within it, down to the unending quest for the ultimate in fine structure—is the domain of quantum motion. That motion, represented in our visual model by a few rather striking conventions, is both the most surprising of all the remarkable discoveries of physics and the one that most resembles everyday life. For it is not the almanac that governs much everyday experience, nor is it the more or less accidental orbits of gravitating systems. What we find is material stability: Like begets like; gold always glitters; bread is nourishing. If atoms were so many tiny solar systems, that stability would be impossible. No two planets are alike; yet all electrons are identical, and so are all atoms, taken species by species. No two stars can have identical corteges of planets, even if we discover plenty of solar systems where until now we have seen none. Detailed study of atomic structure and atomic bonding have shown, in analogy to the solar system, that there is a central heavy bonding center with an external coterie of circling electrons, analogues of the planets. Even the Keplerian laws of motion—in a modified form, to be sure—were found true. The atomic forces are enormously stronger, particle for particle; they proved to be electrical, and thus saturable, not insatiable like gravity. The atom wants to gain or lose at most only a few more electrons. But the sun will pick up additional mass whenever it is offered. Gravity always attracts, but electric forces can repel.

The key is the modular, precisely repetitive, stable form. No large system is like that. Identical building blocks—identical and stable forms kept at an energy minimum and resisting any small disturbance—characterize the atom. They are the signs of quantum motion. The electrons bind to an attracting charge just as they would in large-scale physics; the forces are the same, and the fall from orbit inward to hug the center is prevented by motion. But there the analogy runs out. The quantum motion is of fixed pattern; only certain arrangements in space are open to the moving electrons, and of these only one kind of structure has the lowest energy. Given that state, the electron is present, but it is

This series of spectral lines in the visible light from an electrical discharge tube containing hydrogen stretches from red at the right to the near ultraviolet at the left. The measured spacings obey so striking a simple rule that they could do for Bohr what Kepler's discovery had done for Newton: signal a simple law of force within the atom, as long before was recognized around the sun.

untrackable from moment to moment. No orbit can be traced step by step; there is only an entire stable pattern. Can an electron never be followed along a circular orbit? It can, but only if given an energy much greater than the minimum possible, denying it a larger volume in which to spread out. The energy present is then by no means sharply defined but spread among many energy states, as no stable atom could allow.

Scale determines. Given a larger particle than an electron (say a dust mote, or even an electron with much more energy, like those that paint the TV image), the motion can be followed as sequentially as any circuit of the moon. Quantum motion fully reduces to that of Newton once given the scale conditions of the macroscopic world. But in the atomic regime, no stopwatch, no light flash, no fast camera is adequate to note the electron in its flight between two points within its usual domain; and the use of any radiation or any material sensors that could function on the appropriate scales of time and space would require energy transfers so great as to break up the atom. You might track your electron, but no longer would it be an electron within a hydrogen atom bonded normally to the carbon atom.

For that reason, our visual representation of quantum motion conceals the particles involved; the individual electrons in the bonded atoms are not drawn. Rather, we show the electron charge cloud that they paint out over time, in a pattern that, on the average, becomes sure but cannot be found to unfold from moment to moment. The convention seems a fair one; the concepts it represents are profound, even paradoxical. But in a way they are not so strange as the haphazard quality of the might-be-anywhere orbits of the large-scale world. We are used to a stable world of substances with fixed if manifold properties; individual atoms in a way have no history. They are identical. A gold coin may contain last year's gold from the mines of the Rand or gold won from the California placers in 1848. The two metals are one in kind. All chemical change is by the reassortment of atoms and electrons, never by their wearing out or their gradual readjustment to some long history of attrition. When atoms heal, their recovery is total. Even nuclear decay is ordered and modular; aging uranium always turns into lead atoms of the right sort at the right rate. It is this stability even during change that marks the world of quantum motion, the world of modular identity.

HUMAN SCALE

Between the galaxies and the atoms, the journey pages of this book dwell for a dozen powers of ten nearer the human scale, amidst color, life, and familiar landscapes. This is the realm recognizable in history. It is better known than that of the universe at large and not so strange, in its subtle unfolding, as the atomic patterns, which seem to obey immutable laws whose history we can hardly conceive. Life too has had a long history: evolution. Its intricate mechanisms and its wide adaptations, its fitness like its beauty, are slowly ripened fruits of time, time long enough for Darwinian editing of the

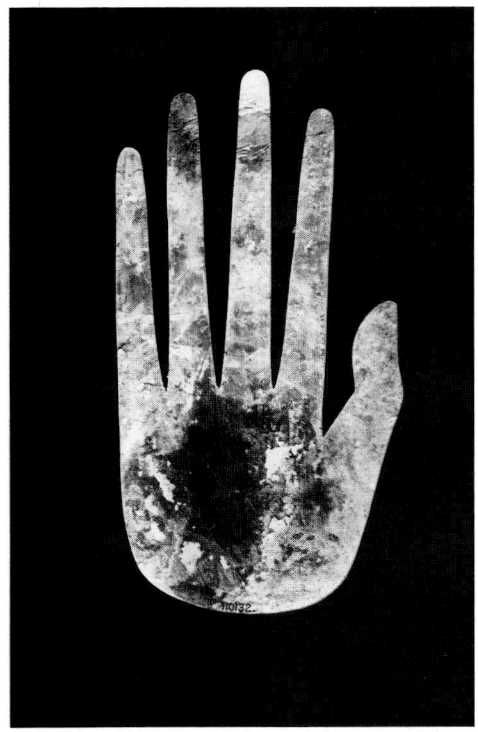

A hand cut from a sheet of mica by an artist of the Hopewell people in Ohio a thousand years ago.

manifold rearrangements of an inner, modular, chemical world. Human artifacts, from wide fields and great cities to tiny electronic chips, undergo a similar though far more rapid pattern of evolution; their changes are entangled with those of the human mind.

This book is itself such an artifact. Held within its time, it is a product of the collective work of many minds and hands: many that have touched it on its way to an audience, and even more whose concepts and techniques it employs, though they are far from the book itself. Its text is expressed in a language we all share; its images are the work of a host of image makers varied in subject and in technique. It is fair to see in the book two distinct streams: one a strict imaginary journey through the universe, the other stream one that includes pauses to look around, to examine detail, to inquire about arguments and conventions, above all to take a direct and sometimes skeptical look at the pieces of evidence so artfully assembled.

The images here, both the pictorial images and the mental structures they evoke, are in part formed by the world as it is. But in some part they transmit an illusion held within human science and human art. That is the best we can do today. Tomorrow the view will differ; we hope it will be more penetrating, more inclusive, freer of misconception, and more beautiful. Meanwhile, here it is: a partly documented version of the cosmos we see in our time. We affirm this day what Plato said so long ago: To write of such admirable structures but to present no visual model would be labor spent in vain.

THE JOURNEY

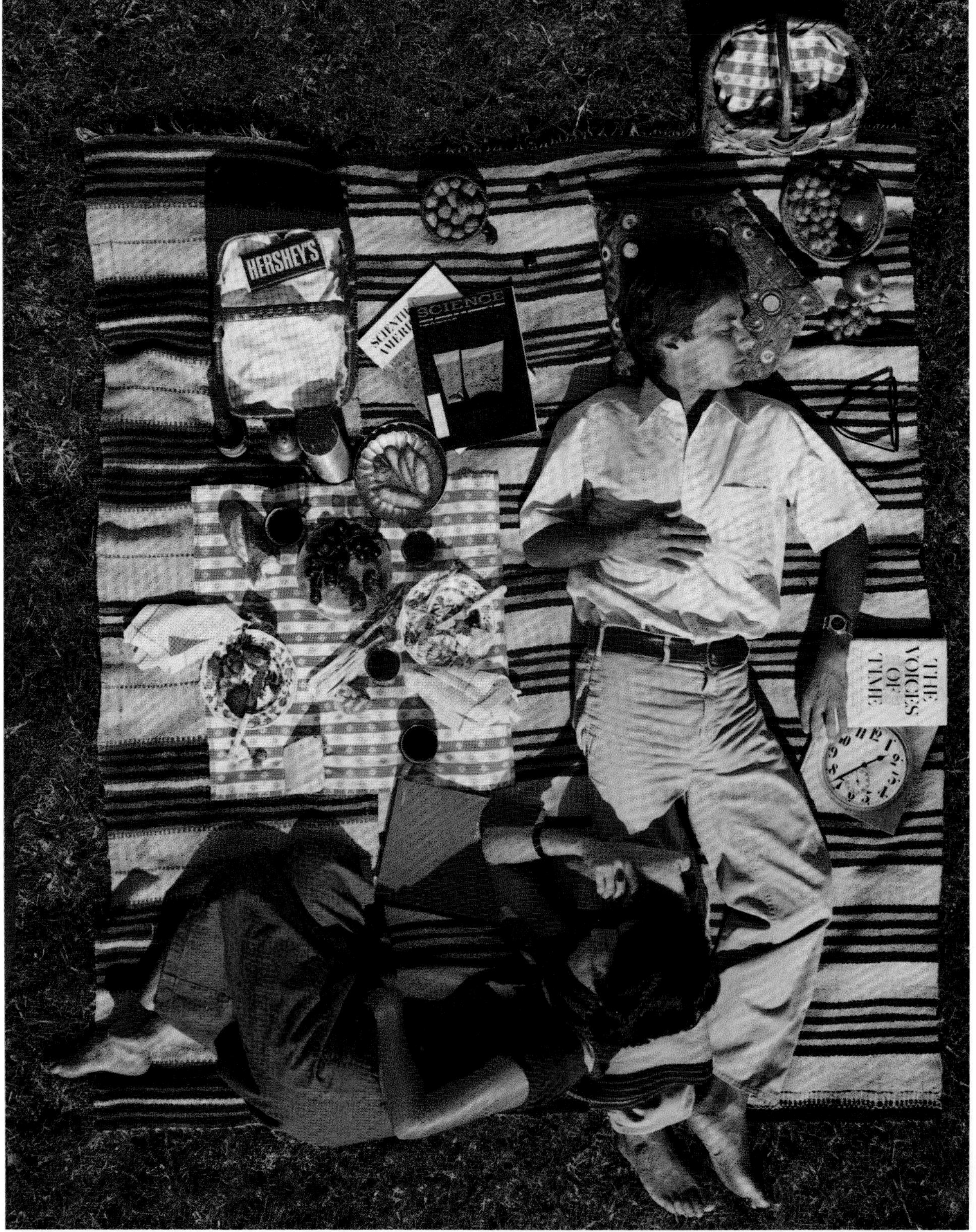
HERSHEY'S
SCIENCE
THE VOICES OF TIME

ADVICE TO THE READER

The core of this book is the scenes on the forty-two right-hand pages that follow. By themselves, they present a visual model of our current knowledge of the universe, showing along one straight line both the large and the small. Each image stands against a black background, a little reminiscent of a darkened theater. Across from every black-framed page is a page of text and picture, a pause at each step along the journey to examine detail, evidence, or the history of our knowledge.

The step from one scene to its neighbor is always made a tenfold change: The edge of each square represents a length ten times longer or shorter than that of its two neighbors. The small central square frames the scene next inward.

The reader is master of the voyage. Starting point and direction are open to choice. The first venture might most happily begin at the familiar, the picnic scene labeled one meter, for the meter is the unit to which all the measurements relate. Then in one direction the pages open to the macrocosm, past the sun and the stars to the darkness of the universe, and in the other direction down from the familiar into the microcosm of living cells and beyond to the incessant atomic motion. The journey can be taken all in one direction, a zoom from largest to smallest, or quite the other way. Each of these offers its view of the world.

The voyage can be made by successive steps, tenfold larger or smaller at each turn of the page. But the reader can skip, to sample the world, say by thousandfold leaps. Equally, the pages offer a reference frame, a marker for exploration of experience in the domain of astronomy, or of geography, or of biology, or of chemistry. Any physical object can be sought out in its proper place along the journey, and so given an appropriate context.

Together these "powers of ten" present "the relative size of things in the universe and the effect of adding another zero."

On the facing page we look out at the universe on a scale of some grandeur: one billion light-years. The Virgo Supercluster of galaxies remains central but hardly conspicuous. We see many other clusters and superclusters of galaxies strung out in lacy patterns, though none near enough to catch the eye.

The view presented is both deep and old. To carry such an image, the light would have had to travel for a time that spans a considerable part of the whole history of earth, starting on its way so long ago that the only life flourishing on earth was microbial.

Yet in the vast realm of the galaxies not much changes during a billion years. One more step outward, one more tenfolding, would bring us to the time when even the galaxies were new, an epoch we understand little. Before that we detect signs of a profoundly different universe, without stars and without galaxies, a bland domain of uniform hot gas. The starry galaxies and even the quasars rightly belong to astronomy; the whole uniform gaseous universe, present before they arose in all their variety, is the proper subject matter of cosmology, now a modern science though heir to a strong tradition of speculative thought.

The gaseous universe of the past was uniform but not stationary. It had a simple pattern of motion; it remained uniform while it became steadily more dilute by expanding as a whole. Neighboring points drift only slowly apart; distant ones move more rapidly in strict proportion; thereby uniformity is maintained. The galaxies retain the old regularity of motion, but their gravitation has since imposed more modest and less orderly motions of rotation and orbital flight: All the distances between clusters of galaxies—but not the bound clusters themselves—still steadily expand.

The obscure feature marked in photo *1* is just about the farthest galaxy ever seen, five or ten billion light-years away. The photograph was made at the telescope in 1979; attention had earlier been called to the faint visible object because it is also a radio source. It is flying away from us—or we from it—at about 0.6 the speed of light.

1

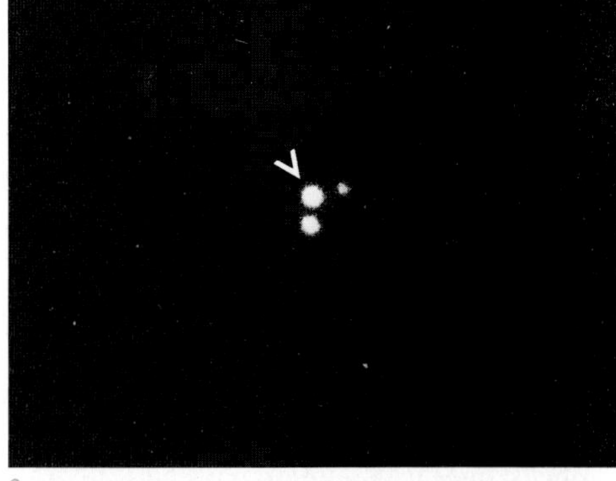

2

The only objects photographed so far at distances greater than that of the faint galaxy in *1* belong to a very distant class of powerful compact sources of radiation, the quasars. The marked spot in *2* is one such quasar, also first picked out because of its unusual radio emission. The image near it is that of a foreground star within our own galaxy; the two appear of about the same brightness, although the quasar is about a million times farther away! It seems now that quasars are especially luminous massive hearts of certain galaxies.

3

A group of faint images *3* is all we can make out of a far cluster of galaxies, distant two to four billion light-years. We detect only the brightest members of this cluster.

10^{25} meters

~1 billion light-years

Most of space looks as empty as this, the glow of distant galaxies like clotted dust. This emptiness is normal; our own bright home-world is the exception. A tenfold larger view would show no new structure, no new void; the universe is roughly uniform at such dimensions. Novelty on so grand a scale is to be sought over time rather than from place to place. All swift change is in the past. This view will dim slowly, for a few billion years at least, as the faint clusters drift still farther apart.

GALAXIES CLUSTER

1 10^5 ly

Galaxies cluster under gravity into groups large and small. This photo *1* shows the central region of the nearest great pool of galaxies, called the Virgo Cluster, as viewed from earth. On the facing page that same Virgo Cluster is represented as it looks seen from a position out beyond it, looking back to earth. At this distance it appears within the small central square: that area will entirely fill the next inward view along our journey.

A couple of thousand galaxies of all sorts and sizes make up the Virgo Cluster, circulating in mutual orbit. It is around fifty million light-years away; our Galaxy and its small group of a few dozen others seem only part of an even bigger assembly of galaxies, vague but enduring, known as the Virgo Supercluster.

3 10^5 ly

A pair of galaxies *3* is caught in the slow act of collision. Mutual tidal forces have dragged out long starry arcs. We see here stars and some glowing gas; the collision is lingering, rather than violent, lasting for a geological epoch. During such an interaction, no stars actually touch; rather, the once-ordering gravitational forces become ambiguous. Some outer stars cannot keep orbital allegiance to their own galaxy but are slowly and inexorably swept with their accompanying gas into long new arms, to fall back someday or to fly off for good into the intergalactic void. Such events are not uncommon, especially within crowded clusters of galaxies.

2 10^4 ly

Here *2* is one of the dominant galaxies of the Virgo Cluster, pristine and undisturbed. It is a fine large elliptical galaxy with many rather old and faint stars. The fuzzy points that surround it are globular clusters, each with some hundred thousand stars; our Milky Way sports a hundred or two globulars of its own, spread far around us rather than gathered into that flat disk amidst bright spiral arms.

~100 million light-years

We look toward our distant home in the Milky Way. But we see mostly one large intervening cluster of galaxies, called the Virgo Cluster. Galaxies as a rule associate into orbiting clusters and groups. There is reason to believe that our Milky Way is itself an outlier of the big Virgo Cluster, responsive to its steady gravitational pull: part of a supercluster. Out there beyond the Milky Way is a good-sized volume nearly devoid of noticeable galaxies.

1

2

This pair of galaxies is seen in two different photos, each exposed for about the same time. But photo *1* admits the wider range of color, which maps even the sparse stars far from the central core, and a bridge of stars becomes clear.

Photo *2* is taken with less light, using one selected color range; that color maps the glowing gas, which lies mainly in the spiral arms.

3

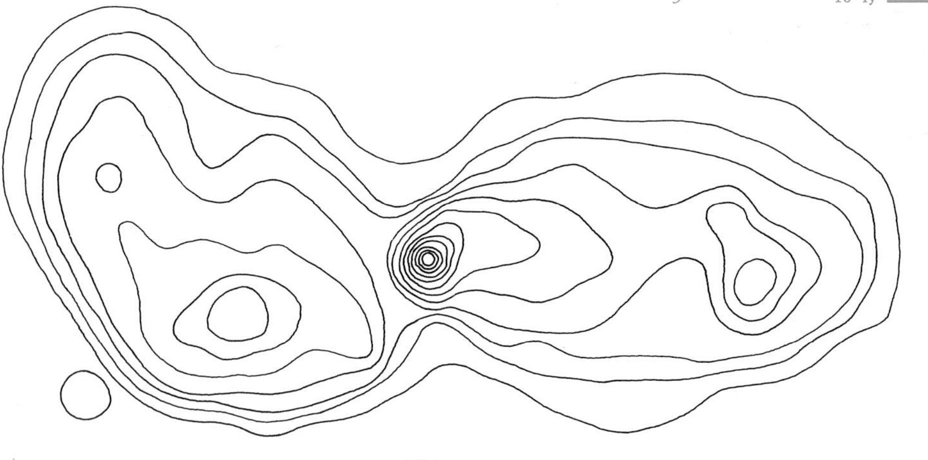

4

The visible form of this unusual galaxy is clearly strange: a dust-belted sphere of stars *3*. The radio telescope discloses it as an enormous radio source: A map of its radio emissions is shown in *4*. Far beyond the central ball of visible stars this galaxy has poured out huge dilute clouds of energy-laden gas that generate copious radio emission but little visible light. The still unseen dust-hidden center, marked by additional x-ray and infrared radiation, must hold some energy emitter far more powerful than any star. Fifteen million light-years away, this galaxy is perhaps the nearest counterpart of that distant and powerful class we call quasars: Its nature will be understood only through inspection by way of many channels.

~10 million light-years ~3 megaparsecs

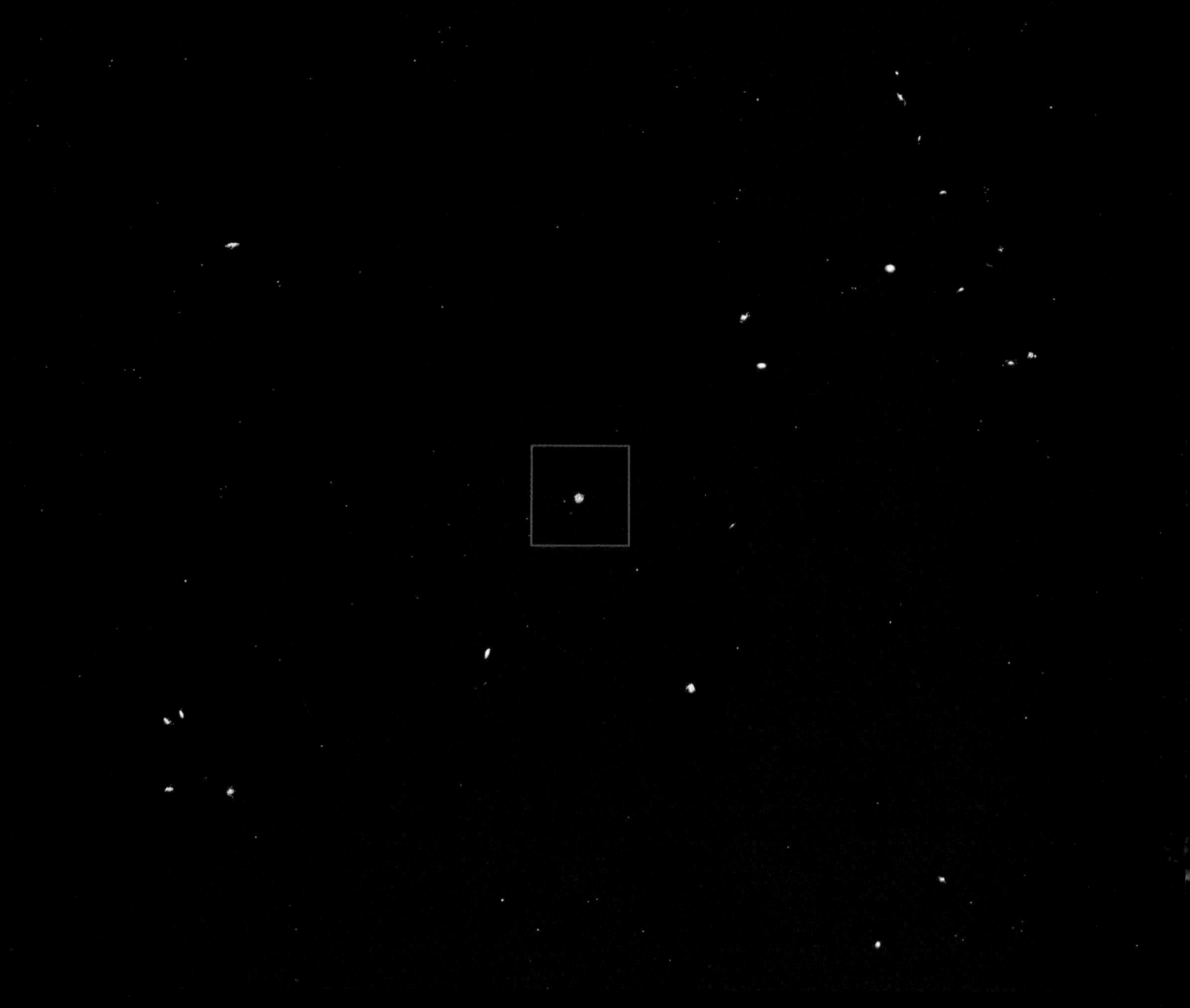

These are the galaxies of our own cosmic region, each single bright spot made by the summed light of stars by the billion. Their mutual gravity binds stars into galaxies, every one a complex swarm of moving stars.

1

25,000 ly

These two irregular companions to our Galaxy–each holds stars by the hundred million—are visible to the unaided eye on dark nights under far southern skies. They look *1* like detached fragments of the Milky Way. Familiar to the peoples of the south, they were first reported to Europe by the Italian navigator Andreas Corsali, who sailed for the Crown of Portugal about 1515, a few years before the first voyage around the world. A drawing published from Corsali's report is shown in *2*. The clouds are named in most languages as fit monument to Ferdinand Magellan, the circumnavigator.

2

The two Clouds are the closest galaxies we can see clearly; they have long meant much to astronomical discovery. But the unaided eye can pick up one other galaxy of the Local Group, a big spiral, strongly resembling our own. (It is best seen on dark August nights high in the northern sky.) This wide-angle photo *3* is specially made to match what the eye sees: The broad band of the Milky Way crowds the foreground, while below it faintly shines the small ellipse of the tilted Andromeda Galaxy, two million light years off. That is the deepest we can reach with eye alone, ten times farther out than the Clouds of Magellan. The same galaxy appears in *4* with the magnification and light grasp of a great telescope: Instruments can bring fuller knowledge.

Any galaxy, large or small, is a swirling pool of very many stars. But the two big neighbors, our Galaxy and Andromeda, share the gravitational task of binding the Local Group of much smaller galaxies.

3

4

50,000 ly

10^{22} meters

~1 million light-years

This flat circular disk is our own Galaxy, the Milky Way, with its spiral structure. It travels in space with two satellite galaxies, the irregular little Clouds of Magellan. Not many galaxies are larger than ours; nor are many seen that are smaller than the Clouds.

A GARDEN OF GALAXIES

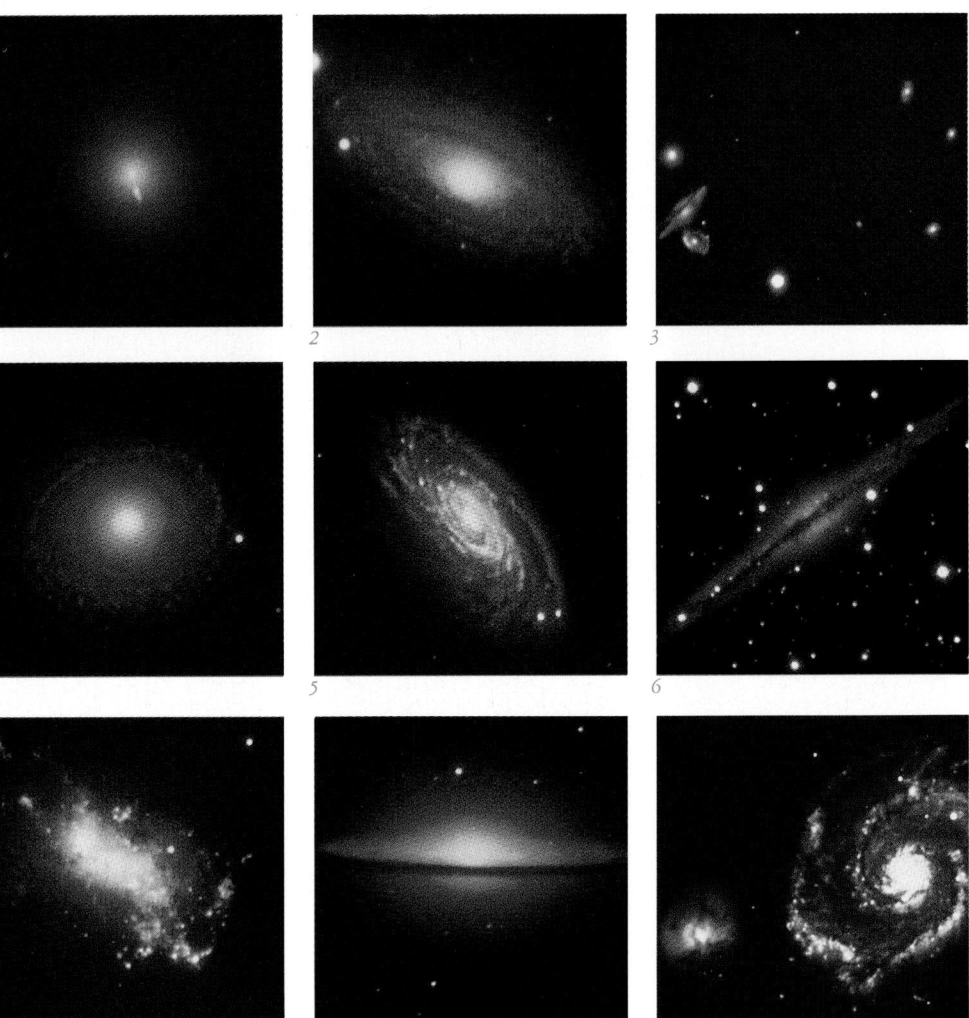

These nine examples serve to suggest the unity and diversity of galaxies. The yellow bulge of old stars is seen very well in *1*, in *2*, and in *8*. Bright blue young stars stand out in *2*, in *5*, and in *7*. The dark lane of dust is striking in *6* and in *8*. In *3* we see a group of seven bright galaxies at the considerable distance of three hundred million light-years; among them they show interesting variety. The three galaxies in *4*, *5*, and *6* are all bright flat spirals, but seen at differing tilts to our direction of view; *4* is face-on and *6* is edgewise. The galaxy in *7* is a small one, not so different from the Magellanic Clouds, dominated by young stars and without marked order. The galaxy in *8* is a big spiral whose central bulge is unusually swollen. The galaxy in *1* is a wonder. Brightest galaxy in the whole Virgo Cluster, it is an elliptical galaxy, a near-featureless sphere of older stars. But at its core is a powerful radio source, and that surprising white jet fingering out from the center is made not of stars but of fast-streaming gas. The pair in *9* are linked in collision; one long arm of a bright open spiral is reaching out in response to the tidal pull of its passing unformed dusty little companion.

Living deep within our Galaxy, unable to travel about it, we have been forced to map it by indirect methods. Our map gains credibility from the overall, if distant, views we have of many external galaxies. Here is a bouquet of galaxies drawn from the few thousand specimens which are all those close enough to see well. Galaxies differ widely, like other complex natural structures, but they share in common varying amounts of three ingredients visible here: a yellowish central bulge of older stars; bluer starstuff, which is the look of bright young stars, often fogged by lanes of dust; and an innermost bright core, sometimes marked by violent bursts of energy.

~100 thousand light-years

We look face-on directly at the Milky Way spiral. A hundred billion stars mutually bound by gravity encircle the central region, some passing close in, some in wider orbits. Our own sun swings with the rest in dignified passage clockwise about the distant galactic center, once every three hundred million years. External galaxies akin to our own are scattered throughout space as far as we can see. They too rotate slowly as they drift.

EXPLODING STARS

"*Mankind is made of star-stuff*," said Harlow Shapley, for the atoms of flesh and bone (save ancient hydrogen) have certainly been built up from lighter ones deep within stars of different kinds. They were spewed out into interstellar space, there to join those gas clouds that long ago condensed under gravity to form the sun and its planets. Most stars shed matter steadily, but the star that ends as a supernova ejects its outer layers with explosive speed. There is good evidence that the heaviest atoms on earth—like lead, gold, and uranium, rare in the universe though salient in human history—are the products of a fierce stellar explosion, perhaps an ancient supernova. That blast itself might have also triggered the infall of gas to form the sun—the fiery death of one star inducing the birth of another.

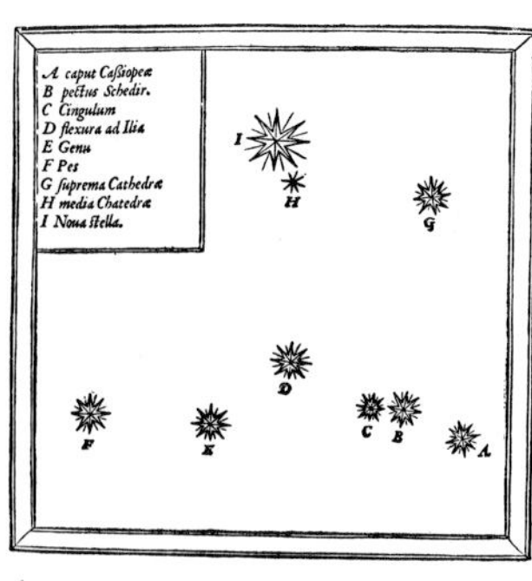

1

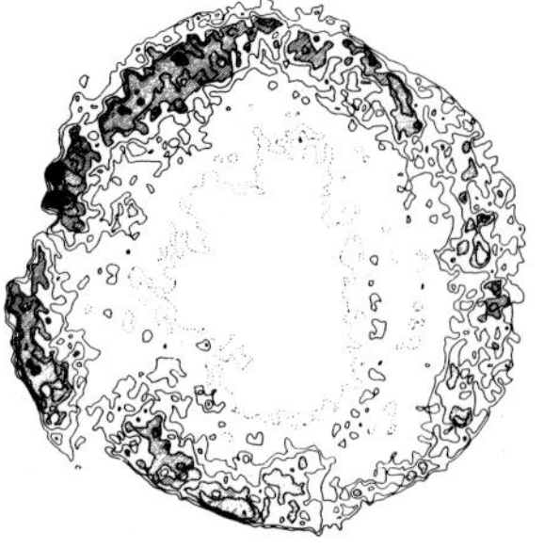

2

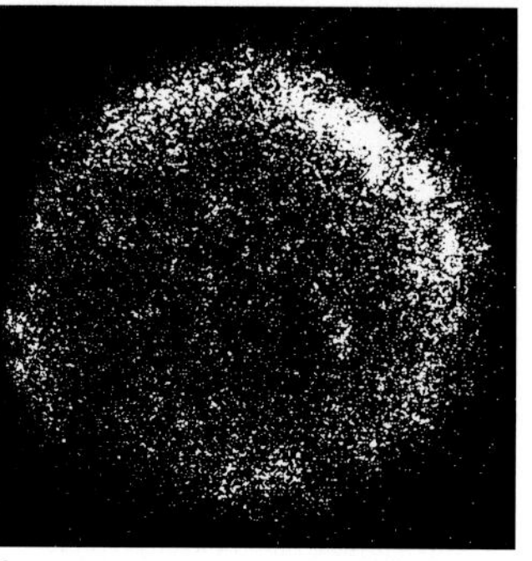

3

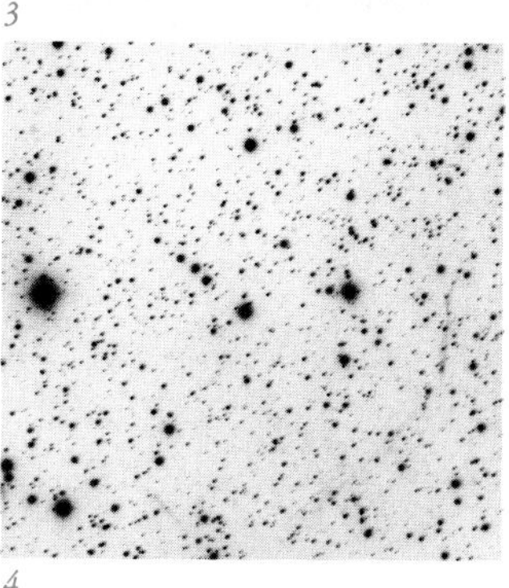

4

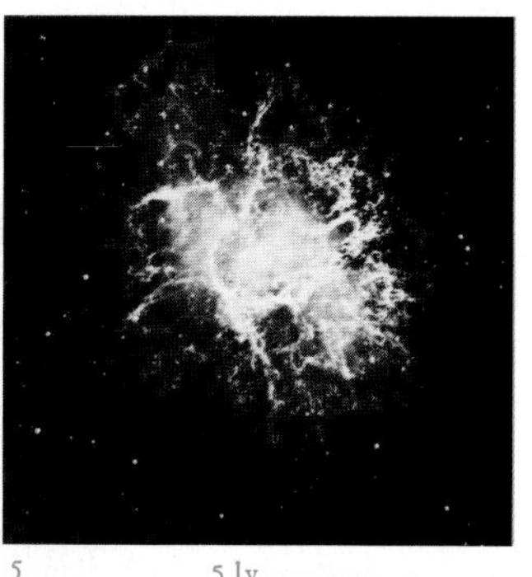

5 5 ly

6

7

In the fall of 1572 many astronomers of Europe watched with wonder a new star that appeared in the constellation Cassiopeia. For weeks it outshone every other star in the sky. Today we call it Tycho's star, for the best account of it—his own sky map of the event is at *1*—was given by the noble young Dane, Tycho Brahe, the finest observer to work without the telescope.

We do not fully understand these prodigious, uncommon stellar explosions, called supernovae. But we see their remnants and can learn much from them. The radio image *2* of Tycho's star reveals a complete shell of radio-emitting electrons around the site of the old explosion. An image caught in the x-ray band *3* confirms the concentric shell of hot matter. Flung out four centuries ago and expanding with an average speed a few percent the speed of light, the shell is distant from us some ten thousand light-years, 10^{20} meters. When we look in visible light at that very place in *4*, we see a crowded field of stars; barely to be made out are the wisps of moving matter that now make up the shell.

Star clerks in the Far East saw a remarkable new bright star appear and fade away in the year +1054, and recorded its sky position. At that very point is this striking remnant *5*, named the Crab Nebula by Lord Rosse, the first to examine it in detail in the 1840s with his large reflecting telescope. Measurable motions confirm that the expansion began about +1100. Out of the expanding cloud pours a flood of radiation of all kinds. Study of this nebula and of the spinning collapsed star-stump within (which still powers it) has been a chief source of our imperfect understanding of supernovae.

No supernova has shown up to be a cynosure within our own Galaxy since the seventeenth century. But each year the telescopes catch about a dozen of them, faint and far off, among the thousands of galaxies we watch. These are two pictures, *6* and *7*, of the same spiral galaxy, M100. The only difference is that in *7* a supernova has appeared, in 1979, at the outer end of the lower arm.

Clouds of stars and glowing gas, with patches of darkening dust, mark the slow-changing spiral patterns of the Galaxy disk. Our distant sun cannot be seen here, but it is in the center of the image, near the border of one spiral arm.

10^{19} meters

BETWEEN THE STARS

At this scale we are squarely in the realm of the stars. The grand pattern of the Galaxy does not show up, though the volume we inspect is large enough to present a million stars to telescopic view. Most of them are faint; only a few are conspicuous enough for the unaided eye.

1

10 ly

2 25 ly

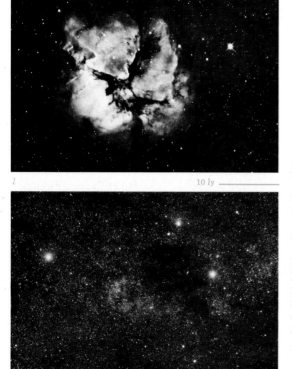

Most of the light in this region is emitted from the surfaces of stars, though there are many clouds of gas and dust out among them that also glow, illuminated by bright stars nearby. Such clouds are then seen as complex bright nebulae, of which one example is shown in *1*. The opposite phenomenon also is found in these big volumes of interstellar space: The light of background stars is here and there obscured by intervening dust, a patchy space fog. Perhaps the most famous example of a concentrated obscuring cloud of dust is shown in *2*.

Stars are widely spaced compared to their physical size. They are distinct, not even close to touching. But photographs tend to blur and merge starlight, for both atmospheric and instrumental reasons. In fact, we might journey on an unwavering straight line right through the disk of our own Galaxy with confidence that we would not strike any star.

The two galaxies of stars shown here are both seen edgewise. The first photo *3*, bordered by a dark foreground blur of trees, shows a portion of the Milky Way as the eye might see it. The other photo *4* is a matched telescopic view, part of an edgewise galaxy some five or ten million light-years away. Their clear similarity demonstrates that here we reside among the stars of a galaxy, peering out this way and that from within its flat star-disk, which appears edgewise as the band of stars we know as the Milky Way. Most of the stars we can see lie within a few thousand light-years' distance at most, well below 10^{20} meters away.

3

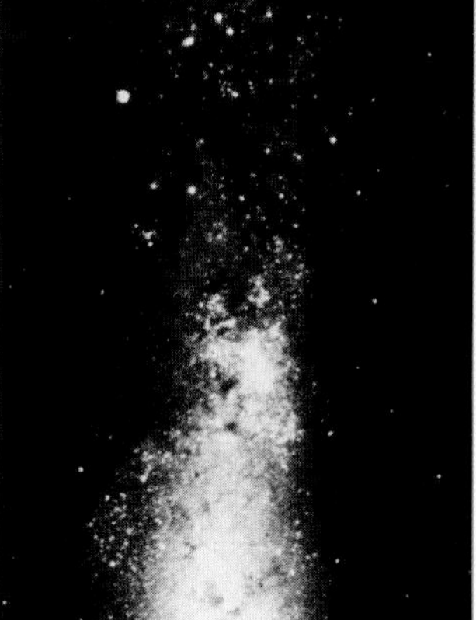

4

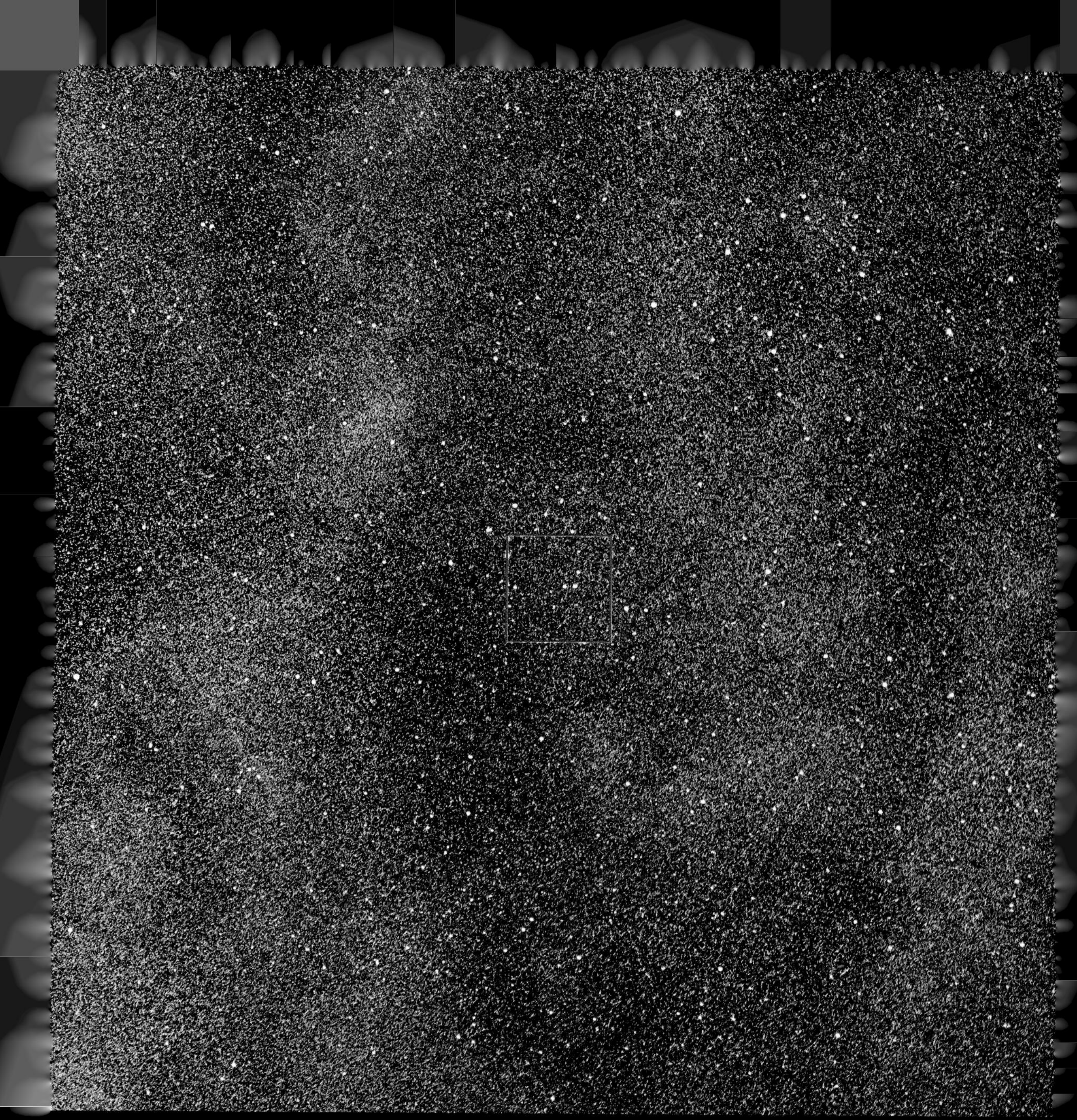

In this view we are within the disk of the Galaxy, right among a host of stars visible here as individuals. Almost every star of the thousand mapped by the old watchers of the sky those who first gathered stars into constellations, lies within this square, our own galactic neighborhood. There are many other stars as well, too faint for the eye to see.

STAR SPECTRA

1

Only a dozen stars among the hundred most easily visible to the naked eye lie around us within a sphere of one hundred light-years' diameter, equal to the edge of this 10^{18}-meter square. One is Arcturus. Two others—the nearest star, the sun, and the second nearest, the Alpha Centauri star system (it is probably a triple star) — are closer, within a sphere one-tenth that diameter. We estimate that about two thousand star systems in all are within the hundred-light-year sphere, most of them still unrecognized faint red stars, hard to find in any way.

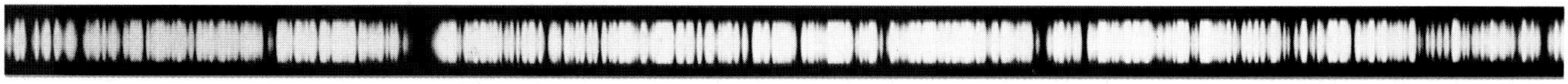

2

A star image is a small blurred point of light in the dark sky. Its light holds a considerable message in code. A hint of its hidden nature is in its color, though color perception is a complex process of eye and brain, and no easy guide. But place a nearly flat big glass wedge as a prism in front of the telescope, and each star image is drawn out as in a rainbow *1*, the differing colors of light bent along slightly differing paths to the photographic plate. At once the stars are given individual character: Their light differs not only in its total intensity, but in the proportions of red and blue.

Below the center of the photo can be made out a strange and wonderful row of half a dozen little rings, like rings of colored smoke. They are the several deviated images of a ring source of light—which is the gaseous shell emerging from the last puff of a distant aging star. The light from that nebular shell does not contain all the colors of the rainbow but only a selected few; between them the plate is dark. This interrupted spectrum is a sign that the source is a tenuous gas, not a compact star.

A second look at the star rainbows may show that they too are not fully continuous bands of color; here and there gaps appear. This effect—a continuous band of color with a few omissions—is rather complementary to the nebular spectrum, where only a few colors show at all. The photo in *2* shows a carefully made spectrum of the light of the star Arcturus. The many narrow interruptions are striking. Here the position alone indicates color, as on a rainbow imaged in black and white. The rich patterns of such color distributions allow inference of the chemical composition, the temperature, and the density of glowing layers of gas. Stars by the hundred thousand have been classified according to their spectral types.

~100 light-years

A skyful of distinct stars: One among them, central, but too faint to pick out, is our sun. The star Arcturus, prominent in the northern sky of earth, shines brightly. Arcturus is intrinsically more luminous than our sun, and here we are nearer to it as well.

The photo shows one field crowded with stars. The stars labeled in *1* are some of the many whose spectrum resembles the sun's. There may be in all a couple of hundred million stars in the Galaxy that similarly approximate our sun in their light output, stability, and lifetime. Are they suns as well to unseen planets, sheltering some long-evolving patterns of life? We do not know.

1

2

Most stars come in pairs, bound together by mutual gravitation. This pair are 200 astronomical units apart. They are so close to us—about 11 light-years—and so well spaced that we easily resolve them. The first stellar distance measured by triangulation was the distance to this pair.

The sun is a representative star. Being a star is a matter of scale: Any sufficiently large and yet not too large globe of gas becomes a star. Such a sphere is held together by its own gravitational pull, but kept from collapse by the balancing internal pressure of its self-heated gas. A star loses heat by radiation—it glows—and must maintain its balance or else contract until it can stabilize. Stars in general become hot enough centrally to release thermonuclear energy by fusion. Once that happens, they can endure a long time, for the nuclear energy available is enormous. A normal star is not an intense emitter of heat for its size; rather, it is so large that even its slow inner metabolism suffices to raise the surface temperature to the glow point: The surface area of a star sphere is small for that great bulk. A normal human releases more heat each second than does the same mass of starstuff. But stars are self-sustaining. The sun feeds upon itself—its inner fire needs neither oxygen nor external fuel—and it can do so over a lifespan of ten billion years.

The smallest normal stars glow faintly a dull red; each contains a few tenths the mass of the sun. They last many billions of uneventful years. The largest normal stars, barely stable against breaking apart, outweigh the sun a hundredfold, glow at blue-white heat, and will not persist. The most populous of the stars, though hard to find at any distance, are those dull red dwarfs. Very few but highly conspicuous afar are the massive, dazzling, young blue stars, which spend their energy stores in a million-year fling.

Most stars are multiple, two or three siblings swinging round one another, in orbits comparable to those of the planets of the solar system. Stars single like the sun are perhaps one or two among ten. Our sun has no starry companions; it has instead its planets. We do not know whether or not any other solar systems exist, but we suspect they may be numerous.

Stars begin as condensations out of the galactic gas and dust. As they shine they often boil out to space. Some end explosively, spewing everything back into the medium to be reworked. All the chemical elements beyond the very lightest ones were made in some star either steadily or explosively by a variety of processes. The cycle is not fully known; we understand quite well the usual star in its prime of life, but the beginnings, the ends, and the many unusual star careers are still at the edge of our researches.

~10 light-years ~3 parsecs

Most of the matter we know is formed into stars, spheres of gas nourished by central nuclear fires that often maintain the glow for a very long time. At this point in the journey, with no star nearby, we see the realm of the stars chiefly as a distant background, no different from the night sky we view from earth. For several frames the star background remains unchanged: The visible stars are strewn so deep in space that these steps are small in comparison. Hence they cause no noticeable shifts.

1

OUR SUN IS A STAR

The largest structure of our solar system is a cloud of comets that dwell in a fuzzy spherical volume around the sun *1*, big enough so objects much beyond fall under the gravitational spell of another star. There the comets, dusty, icy, tailless balls of frozen gases, circulate for ages in giant slow orbits.

The solar system is nowhere to be seen in the 10^{16}-meter view, except for its sovereign, the sun. The orbits of the planets are too small to see. From here the sun is about the distance that light, moving at the ultimate speed, can traverse in one year. It is worth recalling that light speeds from earth to moon in one second plus, and crosses the whole earth–sun distance, one astronomical unit, in just 500 seconds. Even far Pluto is only hours away by light express. A light-year is a heroic unit, fit for the vast spaces between the stars; the solar system is too small for its easy use.

We call the stars "fixed" because they do not appear to change their relative positions at all; the cycles of day and year carry them around all together. But that perception is only approximate. Stars do have their individual motions, although hard to see. One telescopic faint red star, called Barnard's star, moves across the sky faster than any other star known. Its sky motion is minute; if Tycho Brahe, the best of all naked-eye observers, could have seen it, even he might have detected any change in position only after a vigil of five or ten years. The prize of the demanding art of fine positional measurement turns out to be the distances to the stars.

The photo shows four images of Barnard's star *2* along with a brighter reference star. This print is the enlarged superposition of four photographs taken many months apart. The dots show Barnard's star at the four times; all the images of the reference star were fitted one on top of the other. The size of the relative shift is only about a half a millimeter on the plate from the big telescope, yet no other star shifts so rapidly.

The outcome is something of a surprise, which the diagram *3* may help to clarify. Barnard's star moves in a delicately wavy path across the sky. Its side-to-side wobble displays a yearly pattern, a telltale sign. It is the yearly circuit around the sun of the observer on earth, projected onto the smooth flight of Barnard's star. The same repetition appears in the sky path of Halley's comet (see the 10^{14}-meter page). The comet's big loops against the sky are grossly visible; for the star, a hundred thousand times farther away, the effect lies at the margin of our abilities to perceive. This parallax, as it is called, fixes the distance to Barnard's star: nearly 4×10^5 astronomical units, or six light-years. Thousands of the nearer stars have been ranged in this way; it is a surveyor's reliable method, certain in principle if touchy in practice. Stellar parallax has been the foundation of the cosmic scales of distance since its first successes, even before photography, around 1840. There are reports that a microscopic second wobble can be seen in the path of Barnard's star, the pull of an invisible planet orbiting the star itself. But that result has not been confirmed.

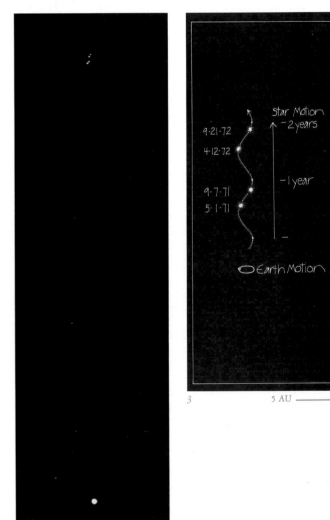

2

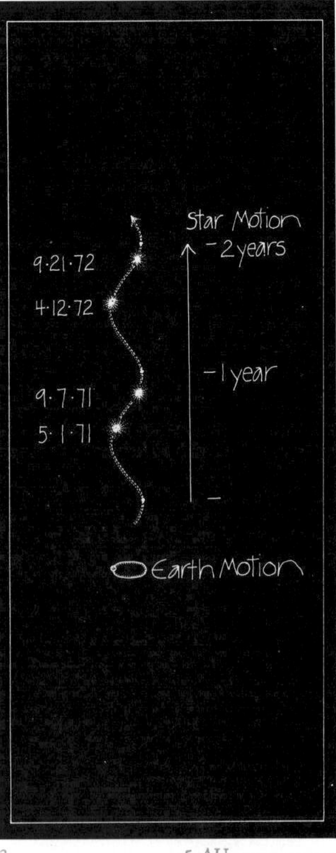

3

5 AU

10^{16} meters

~1 light-year 10 trillion kilometers

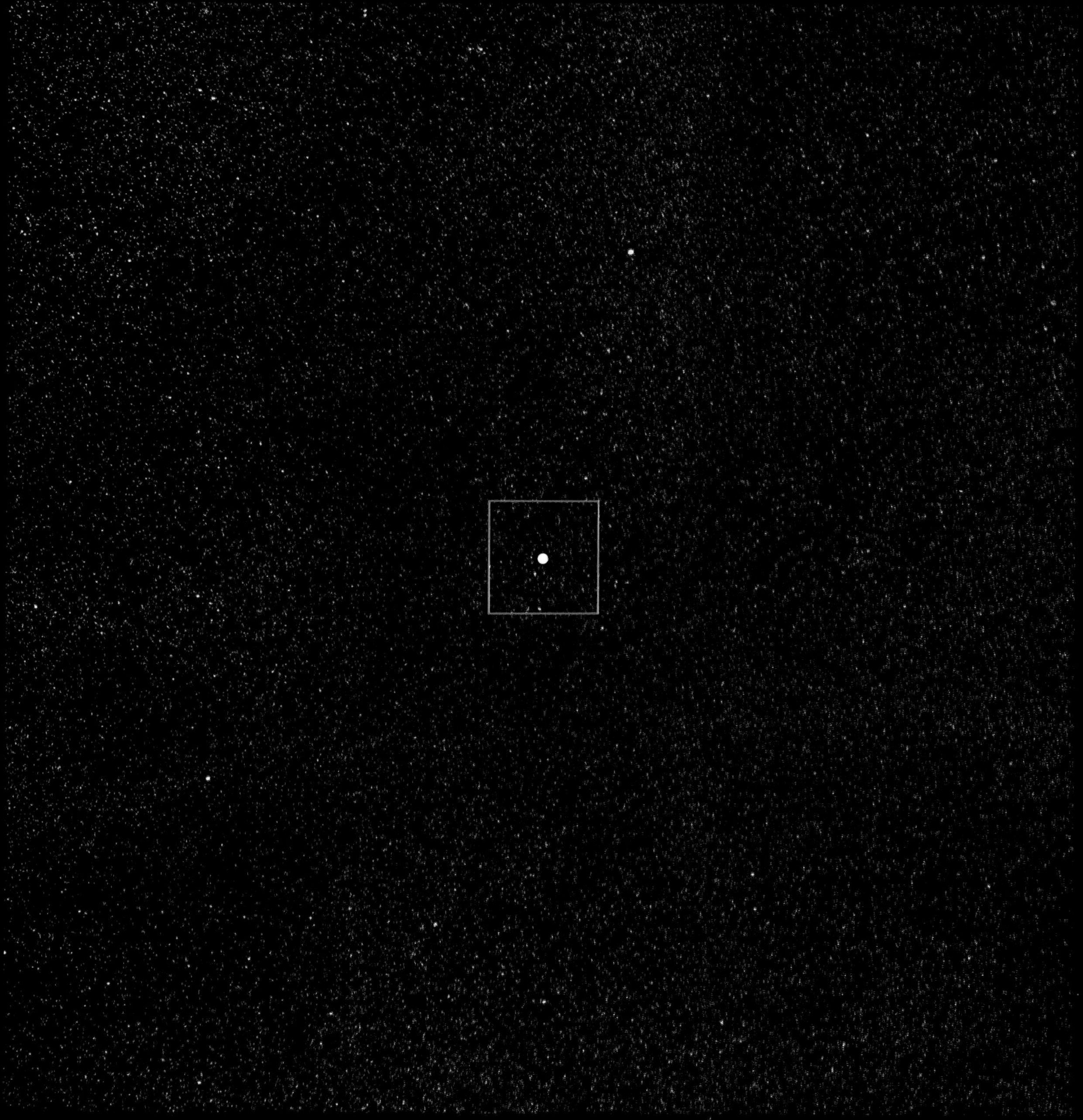

Here one central star is brighter than the rest, only because it is so much nearer. That star is the sun. The contrast between night and day, between the cold glitter of the starry sky and life-giving warmth, is the consequence simply of our planet's location next to one modest star. Once we have drawn away from the sun, we can recognize that it is one star among many stars, and all distant stars are in some way suns.

10^{15} meters

PATTERNS OF STARS

In the 10^{15}-meter square is a star background. It is probably unfamiliar, for it is what a watcher would see from earth looking up not from noonday Chicago but out the other way, into the night sky beyond Australia. Our 10^{15}-meter trip makes no perceptible difference, for all the stars in sight are strewn deep into the background, thousands of times that far away. Some offset might, with effort, be measured, but none that could be noticed with the unaided eye.

Even those who know earth's southern skies well could not point out much of interest, for it happens that this part of the sky sphere is inconspicuous to the eye. These star patterns have been named only recently, mainly for the sake of completeness.

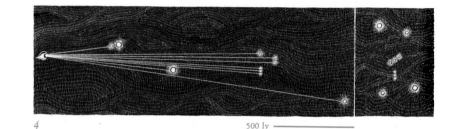

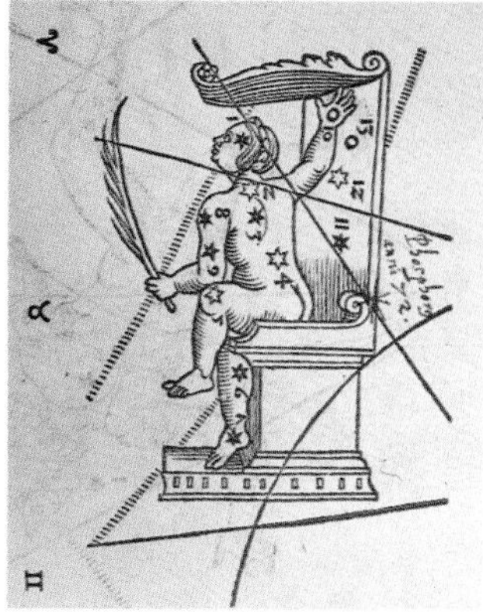

1

2

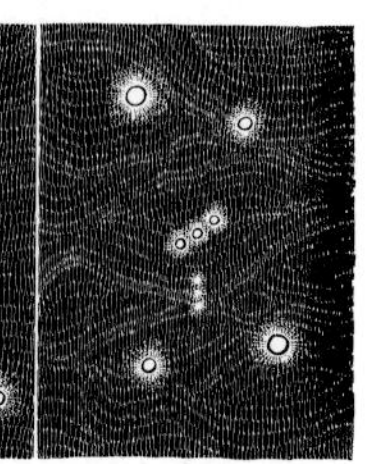

4

500 ly

Three drawings serve as reminders that some constellations have been admired for a long time. The woodcut of Cassiopeia, the nude princess with the feather *1*, is from a sixteenth-century illustrated edition of a Roman collection of sky legends; it was a best-seller for the early printers, in countless editions. The two droll bears *2* come from a popular textbook of cosmography (note the Pointers) printed in 1540. The Chinese Orion map *3* is a portion of a page long preserved in a Buddhist monastery; the manuscript dates to about 940.

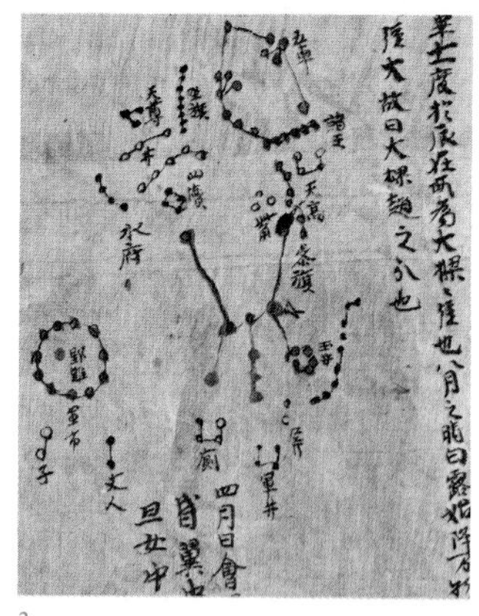

3

Constellations are an ancient part of human culture, but they are hardly structures of nature. They are accidents of our earthly point of view, patterns which line up as they do only from our own position in space and time. That winter spectacle, Orion, is shown in *4* in two ways: as we see it ordinarily, and as its stars lie in space when viewed from the side.

There are exceptions. The Pleiades, that tight shimmering cluster, is a real physical association of close stars, moving together, entangled in a cloud of gas and dust *5*. But most patterns in the starry sky are slowly shifting coincidences, which we pick out because by eye alone we cannot grasp the depth of three-dimensional star space.

In the drawing of Cassiopeia, someone, probably the owner of the volume at the time, has marked in an ink long turned rusty the position of Tycho's new star where it flared up in 1572 to shine brightly for a year or two before fading.

5

5 ly

1 trillion kilometers

Only the sun is to be seen, against a background of fainter stars beyond. Once that was all we knew of the frontier of the sun's system. We know now that a great cloud of icy comets orbits slowly here, though invisible in the weak sunlight. We see comets only as year after year a few fall into the brighter regions near earth. There we catch sight of them, moving in the sky like temporary planets, the sun's fires boiling out their long faint tails.

COMETS

1 5×10^6 km

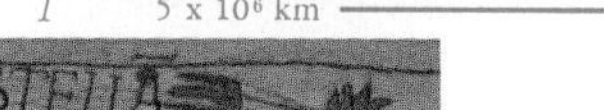

4

2

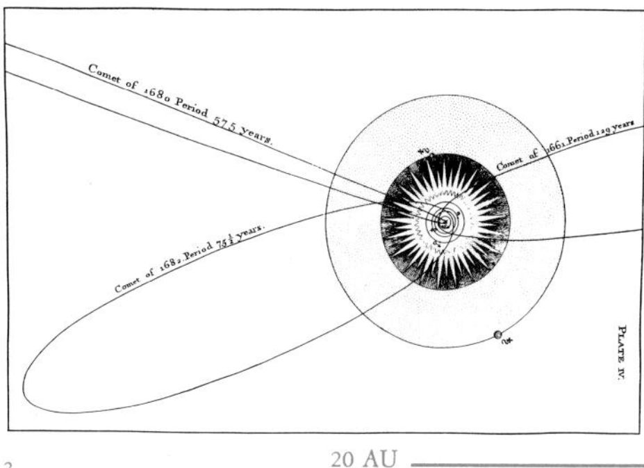

3 20 AU

Once ominous, the strange comets remain unruly subjects of our sun. Unpredicted comets still plunge in from darkness toward new visibility. But some we can confidently announce, like Halley's comet, shown *1* on its last visit in the spring of 1910 with its faint but intricate tail. This comet has been noted repeatedly over history; the Bayeux tapestry, embroidered to celebrate William's triumphant Norman expedition against England, shows it *2* on its visit in the year 1066. Its presence has been documented fifteen hundred years before that. Expect its next close passage in 1986. This is a travel-worn visitor; it will be no flame in the sky. Its wonder is that it will arrive on schedule. The comet's elliptical orbit, stretching out about 35 astronomical units, well beyond Neptune, is in harmony with Newton's theory. In the neat 1750 map of its orbit in *3*, Halley's comet is labeled simply "the comet of 1682."

The path of Halley's comet among the stars for the next few years is shown in *4* as recently calculated. Most of the time the comet is invisible even in the big telescopes. But the path drawn is most instructive. Think of standing magically in space near the sun, your gaze fixed on the stars, no earth spin or orbit to disturb you. The comet plunges in at you faster and faster, its true path a short, smooth arc (look at *3*) until it is almost at hand. The ornate loops of the sky path as seen from the earth by contrast reflect the involuntary motion of all earth observers, whose telescopes are carried to and fro annually as the earth orbits the sun. At first, the apparent path makes only small loops, for the comet is still far off; the loops appear larger and larger as the distance shrinks, until finally the comet passes over our shoulder to head toward the sun.

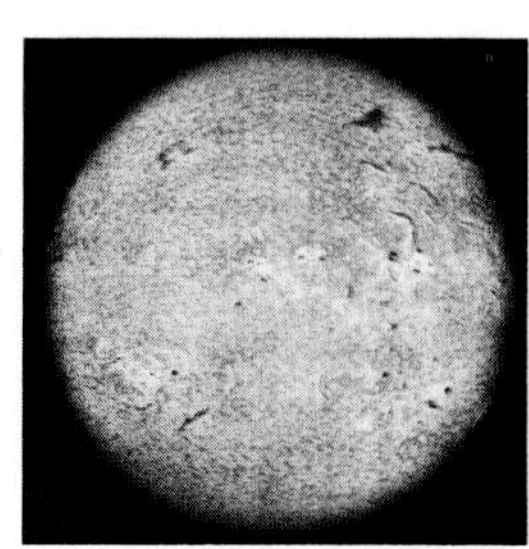

5 6×10^8 m

Stars are not points but hot gaseous spheres. For witness we look at the expressive face of the active sun, on which we see the turbulent magnetic weather of one hot stellar surface. The photo in *5* is taken in a very narrow band in the red to reveal contrast.

100 billion kilometers

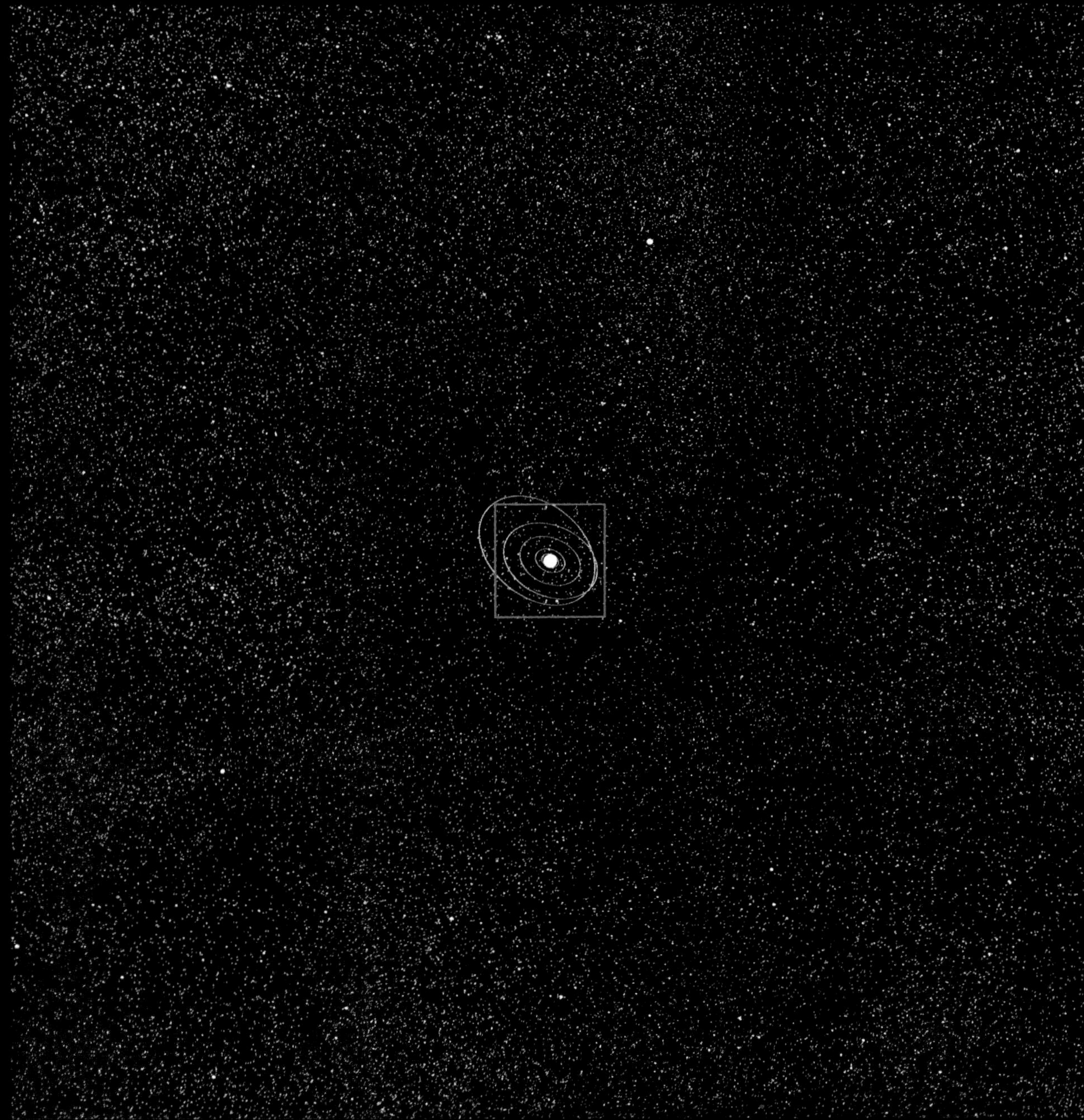

All the sun's planets circulate within the small square. From earth the planets have always stood out, a few strange bright stars restlessly wandering in a skyful of unchanging patterns. Seen here from outside, the planets take on their Copernican aspect; they move around the sun on these nested ellipses, mapped by colored lines.

1

50,000 km

Look from earth outward to any of the big planets, and the face you see is always nearly full. The dreamlike image in *1* is the first in all history to look back at crescent Saturn, viewed from five million kilometers beyond, its shadow black across the famous flat rings. The same robot TV camera on Voyager 1, a few days before it passed forever outside of Saturn's orbit, had ventured through a gap in the rings. There it showed us close-up images like *2,* a thousand subtle rings and ringlets where from earth we had made out only a handful. We know these rings for a remarkably flat procession of icy bodies in orbit, of every size from dust motes to mountainy chunks, in frequent grinding collision. Rings are no longer uniquely Saturnian for us; Jupiter and Uranus bind rings, too, inconspicuous and newly detected.

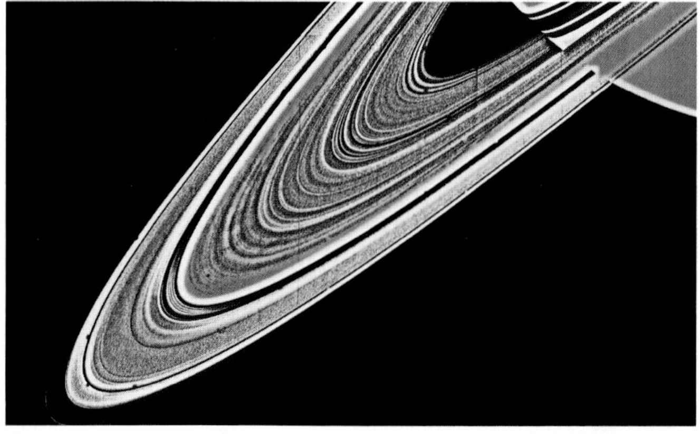

2

5000 km

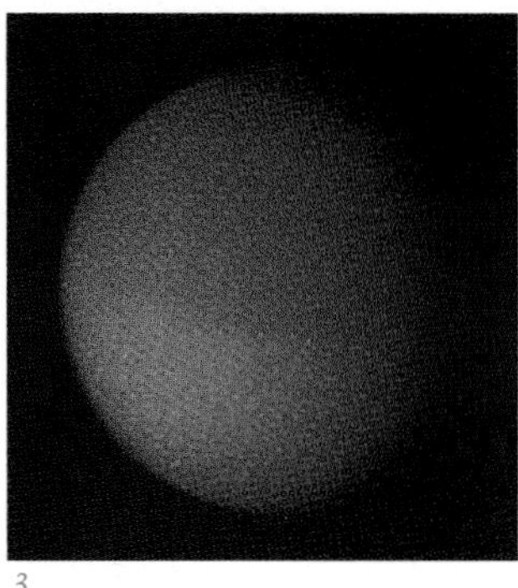

3

4

100 km

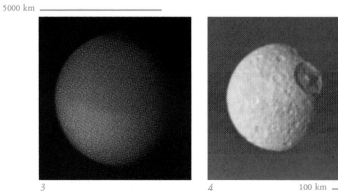

In addition to the myriad objects that make up the rings, we find more and more satellites in isolated orbits around the major planets. Saturn has many moons; they include big ones like Titan *3* and little ones like Mimas *4,* one huge impact crater conspicuous on its many-cratered surface. That old encounter must have threatened to fragment the three-hundred-kilometer ice moon. Mimas is a close companion to Saturn, circling not far outside the outer edge of the rings.

10 billion kilometers

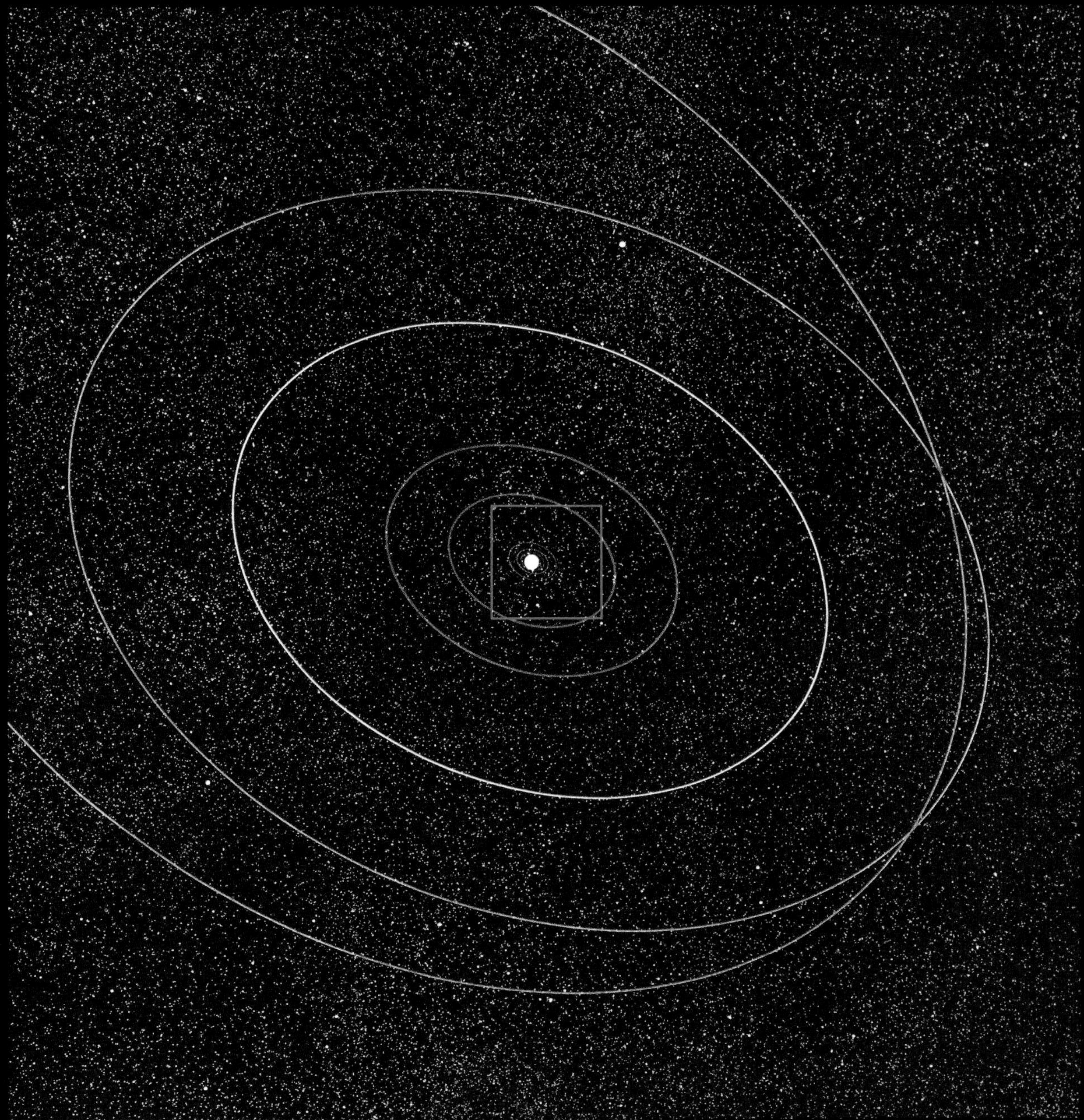

The paths of the outer planets fill this picture. That strongly tilted orbit belongs to little, awry Pluto. The four others are those of big Neptune, Uranus, Saturn, and Jupiter, with their many satellites. Between Jupiter's path and the sun run the inner planets in their smaller orbits. The planets circulate counterclockwise here, all in nearly the same plane, which we view at an angle: The planetary system, apart from Pluto, is flat as a pancake.

1

2 10^5 km

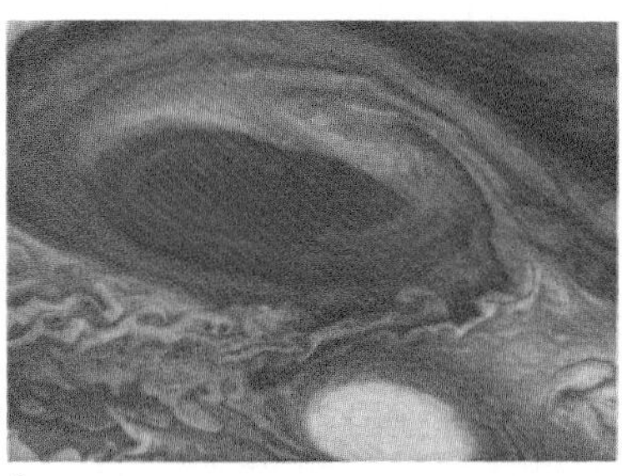

3

Three views of Jupiter, the whirling spots and bands revealing the motions of its thick outer atmosphere colored by enduring clouds of uncertain compositions, show breathtaking visual grasp. The grainiest *1* is what we see with a big telescope from earth, over a long distance, peering first through the disturbed air in which earthly telescopes dwell. The next *2* is fine-grained by comparison, the camera on Voyager 1 now in clear space and only thirty million kilometers from the planet. In the 1980 close-up of *3*, Voyager 1 has gone almost tenfold closer still; the turbulent Jupiter flow is impressive. The Great Red Spot there is an upper-atmosphere pattern—a vast drifting cyclone—that has persisted for centuries, waxing and waning over the years. The smallest distinguishable detail in *3* is about 100 kilometers in diameter.

These three successively more detailed images realize in one way a chief theme of this book.

Io, closest to Jupiter of the four Galilean satellites, is physically extraordinary. Voyager 1 found on it about ten active volcanos —photo in *4* shows the plumes from three of them in eruption at once—that spew out sulfur and its compounds instead of terrestrial ash and lava, the dusty plumes flung 150 kilometers high into space above Io's airless, mottled orange surface.

The moons of Jupiter, four of which compare in size with our own moon, circle swiftly around the planet, as Galileo first made out from a Florentine garden in January of 1610. His manuscript in *5* records the ballet of these moons. Within a couple of years he unraveled their choreography: shadows, eclipses, rhythms. Jupiter's system provides a compact analogue to the sun's whole far-flung system.

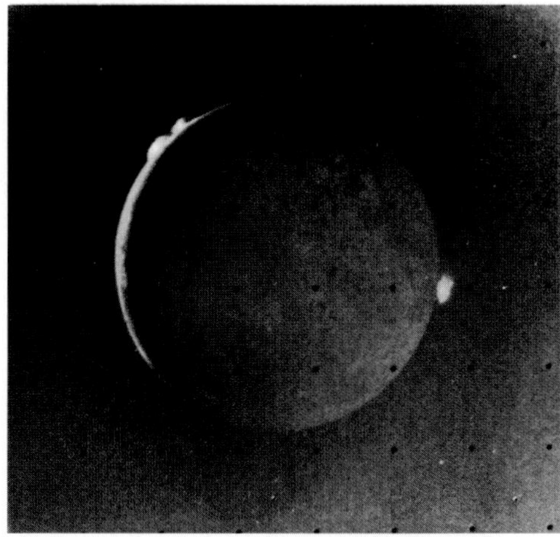

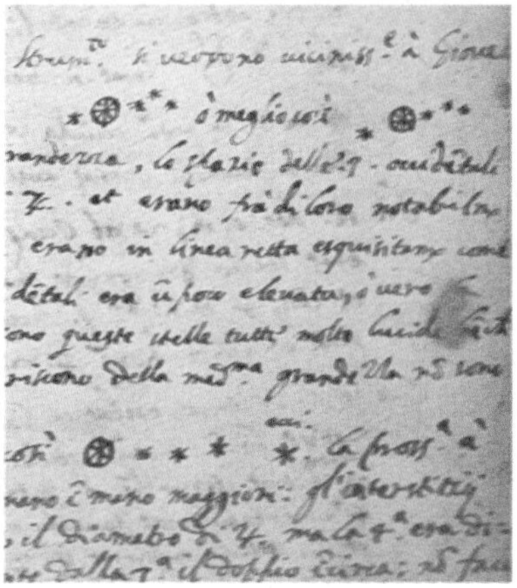

4 10^3 km

5

1 billion kilometers ~7 astronomical units

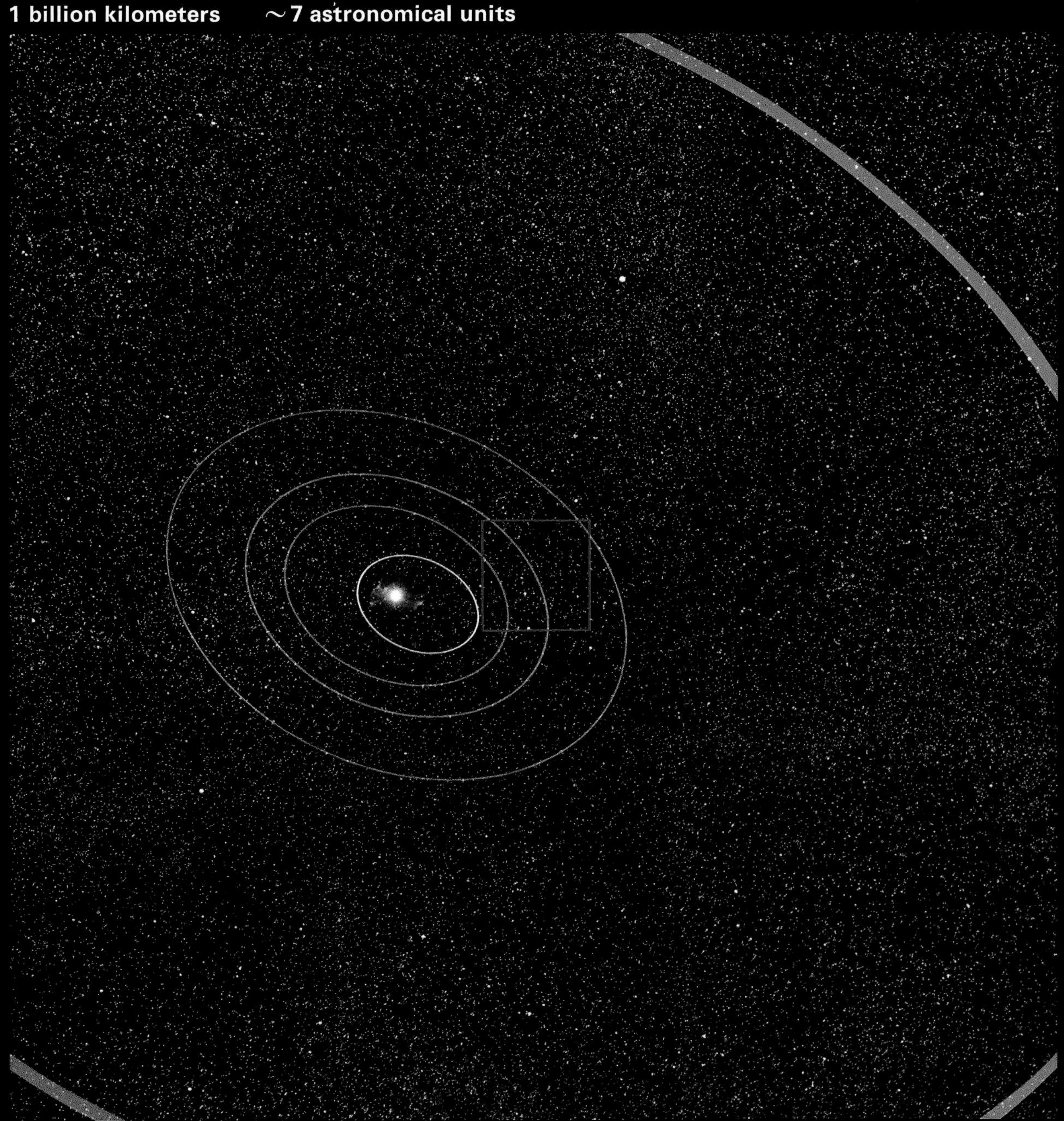

Enclosed in the path of massive Jupiter, these are the orbits of the smaller earthlike inner planets: Mars, Earth, Venus, Mercury. Another swarm of objects too small and faint to make out without telescopic aid is present as well: Asteroids and meteors ply this darkness in the belt between the orbits of Mars and Jupiter.

INNER PLANETS

The inner planets are neighbors and counterparts to earth; predominantly rocky, neither gaseous nor icy, of little bulk compared to the outer gas giants. The planet Mars appears in *1*, as a dawn crescent seen from space, its south pole under night, only the edge of that polar ice cap visible, with a big impact crater brightened by fogs and carbon-dioxide frost. The upper portion of *1* shows one quiescent volcano emerging above the haze, its slopes decorated by a morning cloud forty kilometers high in the Martian stratosphere.

On the Mars surface we are visually at home, even though there is nothing much to breathe in its thin bone-dry atmosphere. Images *2* and *3*, a pair taken on the spot a few months apart by Viking Lander 2, show a desert scene of boulder field and dust—in *3*, frost-whitened in winter. The crisper shadows of the winter's day suggest that the sky was less hazy. Arid though Mars is, hints of liquid water are there—very little currently, much more in the ancient past. In no other place save earth is that vital fluid yet suggested.

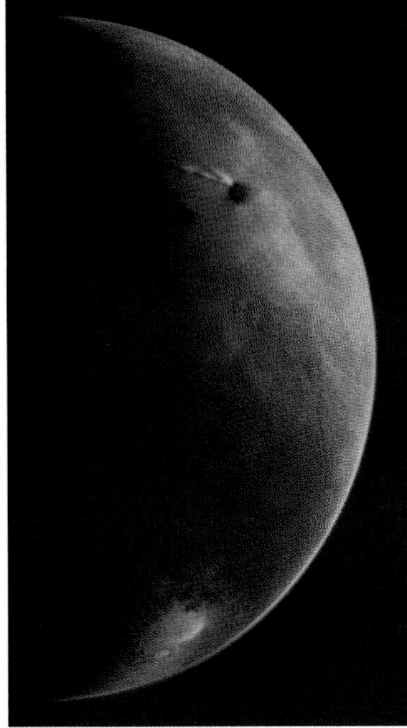

1 2000 km

2

3

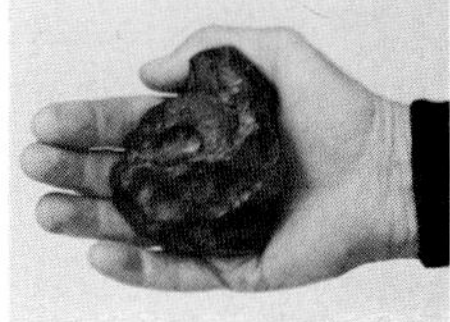

6

This piece of extraterrestrial nickel-iron *6* fell in Arizona about twenty thousand years ago, part of the house-sized body whose vaporizing impact made the famous Meteor Crater. Many thousands of such meteorites, diverse in substance, fallen by chance to earth, have been collected. Some originated within the asteroid belt; others are the catch from one or another drifting, fragmented, eroding comet.

4 10,000 km

Venus too is a world, but its atmosphere is thick and concealing. The image at *4* is a photo taken in ultraviolet; all we can make out is a pattern of drifting cloud—possibly made of droplets of sulfuric acid, surely not water. The hot surface is rocky, as the photo *5* shows, returned by a Soviet probe in the hour before its insulated camera died once the heat of that surface, the temperature of a soldering iron, finally leaked in.

5

Mighty, majestic, and radiant,
You shine brilliantly in the evening,
You brighten the day at dawn,
You stand in the heavens like the sun and the moon,
Your wonders are known both above and below,
To the greatness of the Holy Priestess of Heaven,
To you, Inanna, I sing!

Sumerian hymn to Venus
from a text of −1900

100 million kilometers

Now we see the inner solar system. The green arc is traversed by planet Earth during some six weeks each September and October.

10^{10} meters

THE EARTH'S PATH

Though we had anticipated the scene, real photos of the whole earth fragile in space are marvels still. We had seen earth as it rises, above the horizon of the moon, another demonstration of its planetary nature. But only in 1977 did we first see an image like this one *1*, a fully Copernican perspective of the earth and its companion, twin crescents in space viewed as from another planet. The scene was captured in orbit by the camera of Voyager 1, from a distance of twelve million kilometers, not long after its departure for Jupiter and beyond.

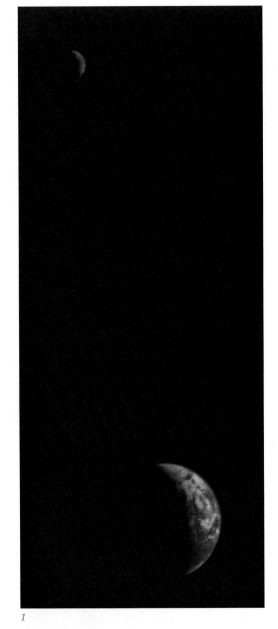

1

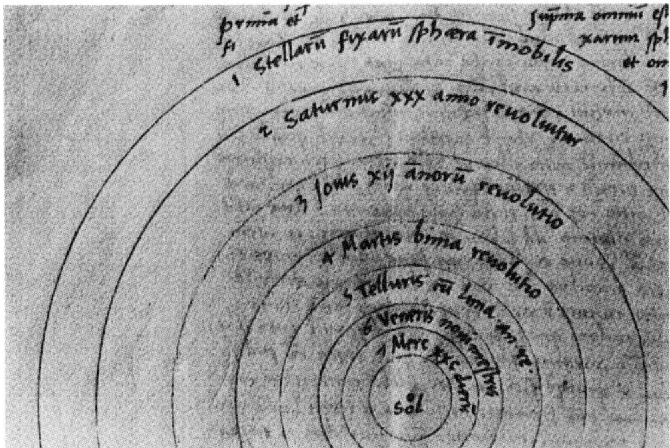

2

During the Age of Discovery, around 1510, not long after Columbus, but before Magellan there circulated privately among a few of the learned a small manuscript by Nicolaus Copernicus, a respected official of the cathedral of Frombork, in Poland. In it he argued that the earth is the center of the moon's path but not of the universe, and that, with all the other planets, it revolves around the greater sun.

Copernicus' proposal was not unprecedented, but it was revolutionary in his time, when the central importance and majestic immobility of our earth had gained theological support. The prize to be won by setting the earth free among the celestial wanderers was nothing less than the unity of the entire cosmos, grasped much later by the seventeenth-century founders of modern science. Galileo at thirty-three wrote to Kepler from Padua in 1597: "*. . . Many years ago I became a convert to the opinions of Copernicus. . . . I have written many reasons . . . which up to now I have preferred not to publish, intimidated by the fortune of our teacher Copernicus who though he will be of immortal fame to some, is yet by an infinite number . . . laughed at and rejected.*"

Copernicus' manuscript *2* for his long-delayed big book, published only in 1543, the year he died, shows his own simplified diagram of the solar system.

10 million kilometers

This path marks the earth's way for four days in October; within it the moon's route is indicated relative to earth. The moon at all times lies somewhere on that small ellipse which moves along with the earth in its orbit.

THE MOON

The moon from her orbit illuminates more than the night. Far enough away so she is plainly celestial, yet near enough so the first of all astronomical telescopes could disclose her earthy nature, her light pales every other star and planet. The moon has given a unique stimulus to science, to art, and to the general imagination. She is a close relative to the earth, but whether she is by her origin daughter, sister, or wife we are not yet sure. The moon could be wrapped neatly in a cloth the size of the two Americas.

A few of us have been there and back.

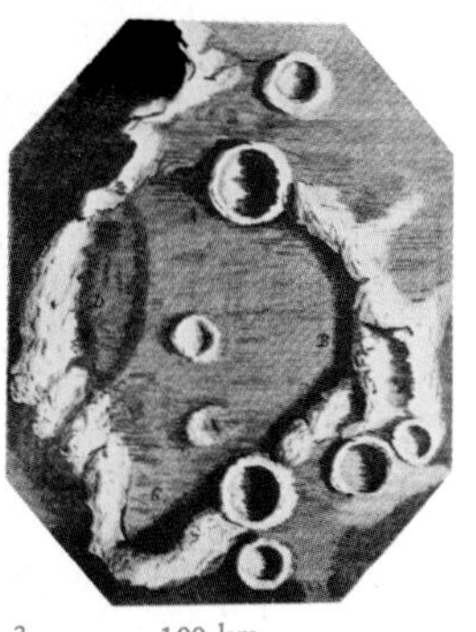

3

In the same year in which William Shakespeare began *The Tempest*, Galileo wrote: "*Almost in the center of the moon there is a cavity . . . perfectly round in shape. . . . It offers the same appearance as would a region like Bohemia if that were enclosed on all sides by very lofty mountains arranged exactly in a circle.*"

The Englishman Robert Hooke drew this detail of the moon *3* through his telescope in 1664.

1

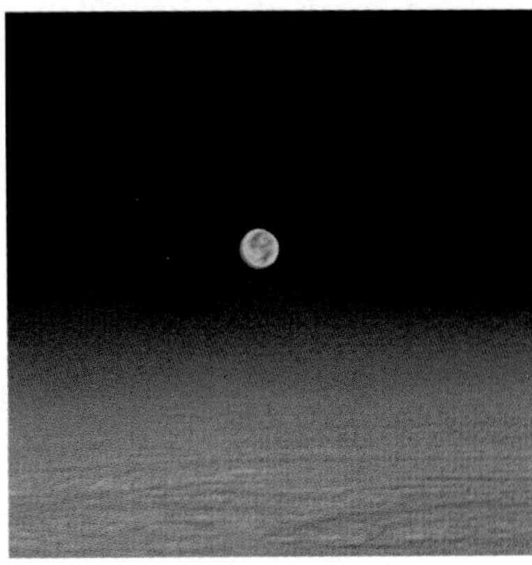

2

The crescent earth *1* rises big beyond the rocky horizon of the far side of the airless moon. The full moon *2* also rises past earth's air-smoothed horizon, viewed from low earth orbit just above the blue of the sky. Because we look across the same distance whether we gaze from earth to moon or from moon to earth, the size comparison is direct.

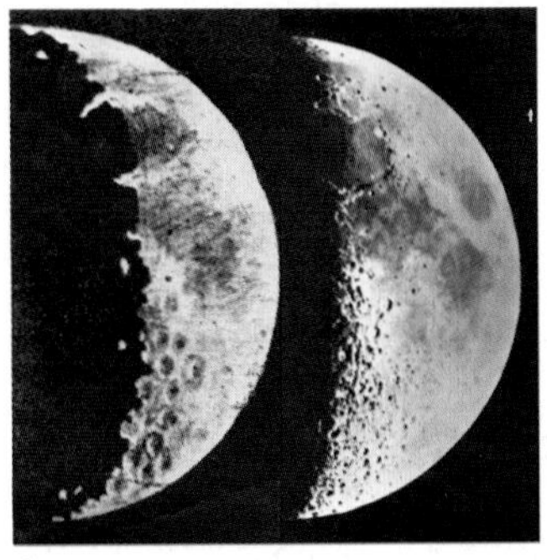

4 *5*

Through one of the first good telescopes he and his expert mechanician had constructed, Galileo saw and drew the magnified moon *4*. Matching his image against a photograph made 350 years later *5*, both showing the crescent moon with its craters along the line of lunar sunrise, offers an idea of how well Galileo understood and reported the first view of the moon as another rocky world.

6

On the trip to the Littrow crater field: Apollo 17, 1972.

7

The impression of a cylinder seal, cut twenty-five or thirty centuries ago in Assyria, celebrates the date palm and the crescent moon *7*.

1 million kilometers

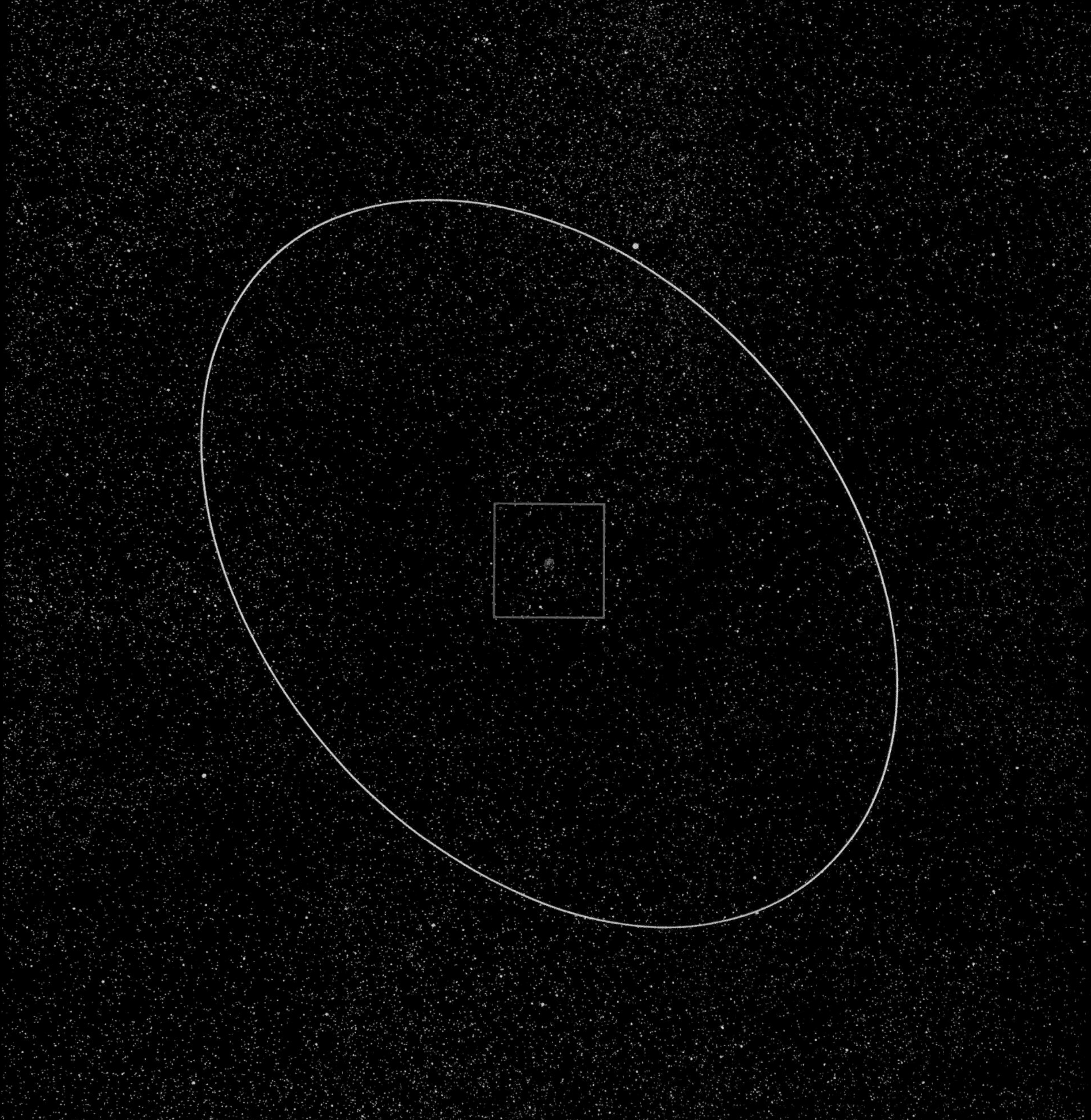

The farthest place our own kind has yet visited is the companion moon, our nearest celestial neighbor. Bright moonlight and the tides witness her proximity.

10^8 meters

THE WHOLE EARTH

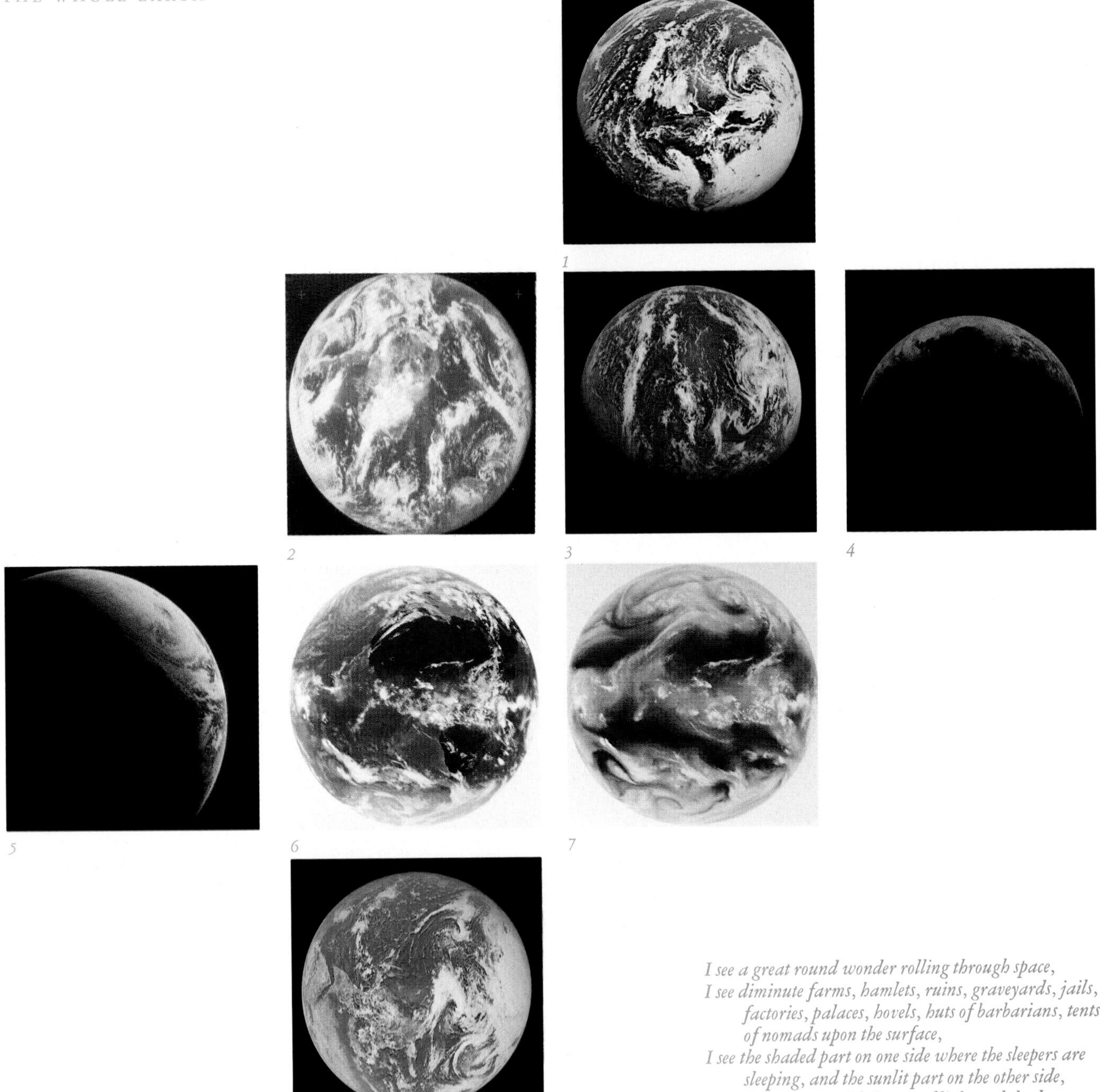

I see a great round wonder rolling through space,
I see diminute farms, hamlets, ruins, graveyards, jails,
factories, palaces, hovels, huts of barbarians, tents
of nomads upon the surface,
I see the shaded part on one side where the sleepers are
sleeping, and the sunlit part on the other side,
I see the curious rapid change of light and shade,
I see distant lands, as real and near to the inhabitants
of them as my land is to me.

Walt Whitman
"Salut au Monde!" 1856

100 thousand kilometers

The whole earth appears, isolated, elegant, and fragile. We recognize our globe in open space, a spacecraft in orbit, no Atlas and no turtles to support it. Its smooth, swift motion around the sun carries it across such a square as this every hour.

10^7 meters

AIR, WATER, EARTH

The whole earth is close up here at 10^7 meters, though it is not quite the whole. That is not mere awkwardness in cropping the photo, but rather represents the underlying structure of the metric system of units. For its originators tried to rest their unit of length on a firmer base than the reach of any king. They defined the meter by setting the distance from pole to equator as measured on a meridian passing through Paris at ten million meters exactly, or ten thousand kilometers. (They were a bit off; the actual distance is about two kilometers more out of the ten thousand.) But the diameter of a sphere with that girth is rather larger than 10^7 meters. A square whose edge is set at the seventh power of ten in meters must cut the sphere.

2

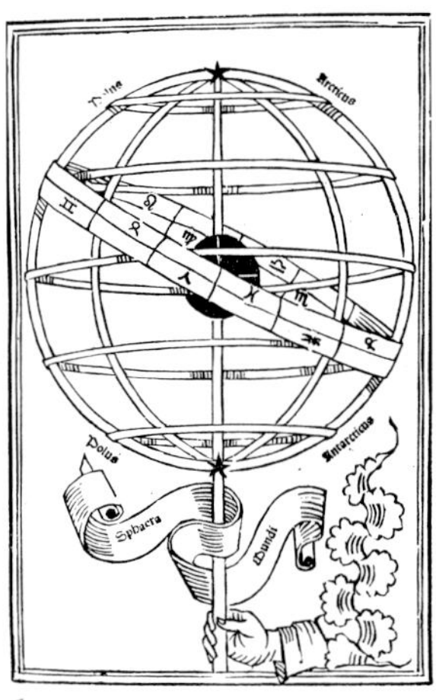

1

It is easy for us in a time of satellites to see the earth as a sphere. But that knowledge, guessed at from symmetry or inferred from shadows and travelers' accounts of strange starry skies, is old. It was old even to the Alexandrians, to Ptolemy himself; it was old to the Arab astronomers; it was old and well known in the medieval universities of Europe; it was old to Christopher Columbus and his backers; it was fully understood by Magellan, who sailed around the world for the first time, a generation after Columbus.

We celebrate here that marvelous ancient understanding, firm before clocks, cameras, radios, and satellites. The illustration in *1* is a 1489 version—just pre-Columbian—of the spherical earth from a classic introductory textbook. First published in Paris about 1220, the book was required reading for astronomy students on many an old university campus for four centuries. The earth is shown quite round within significant celestial circles, all drawn as a real hooped instrument.

Our earth is almost palpable here. It is the blue planet, blue with sky and sea, and a bright one, white with cloud, much brighter than the ash-gray moon. The continents appear clearly when the weather allows; their jigsaw puzzle is evident where the shoulder of South America fits the Gulf of Guinea overseas.

Sky, sea, even land, are thin shells; the bulk is held within that roundness. The sky is infinite in depth to the star-watchers at night, but it is only twenty or thirty kilometers thick by blue day. It curves neatly around the globe, as you can see in *2*, a photo made from Apollo 11 as it headed off for the first moon landing. All the four elements are seen there: The dying storm by sunset over the Indian Ocean west of Australia is a cloud layer planed smooth, the sky curve delicate behind it. Even the land we know is an outer skin; its average height worldwide is a kilometer above sea level, and the oceans average under four kilometers in depth. Only the rocky planet itself is truly three-dimensional. All the world we inhabit and traverse is relatively no thicker than a good coat of paint on a ten-inch library globe.

10 thousand kilometers

The earth in detail: blue sky, white clouds, dark seas, brown lands, a globe turning always eastward. The makers of maps had for three centuries prepared us for this sight, but it became real to eyes as well as to mind only around 1967.

THE GREAT LAKES

For in their interflowing aggregate, these grand fresh-water seas of ours—Erie, and Ontario, and Huron, and Superior, and Michigan—possess an ocean-like expansiveness. . . . They contain round archipelagoes of romantic isles. . . . They mirror the paved capitals of Buffalo and Cleveland, as well as Winnebago villages; they float alike the full-rigged merchant ship, the armed cruiser of the State, the steamer, and the beech canoe. . . . They know what shipwrecks are, for out of sight of land, however inland, they have drowned full many a midnight ship with all its shrieking crew.

Herman Melville,
"The Town-Ho's Story,"
from *Moby Dick*

About forty 10^6-meter squares of this regional size would patch out the entire land area of the New World; the hemisphere is mostly seawater. The paper-thin sheets of the Lakes are in a steady state; they drain to the sea over Niagara's scarp about as fast as rain and melting snows supply fresh water.

Ice carved and first filled the Lakes, scooping and ponding. Mountains are geologically old, but most lakes are new and swift to change. It is only a few tens of thousands of years since different lakes, ancestral to the present chain, lay here. The rich flat lowlands were silted by lake and meltwater, and the looping gravelly rises in the land are deposits by lobes of glacier ice.

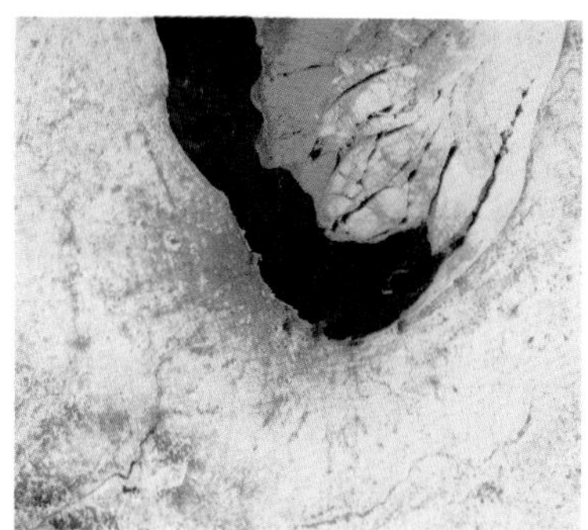

1 100 km

Ice still winters here. The satellite shows the lake *1* from orbit one February: all the land snow-covered, and floating pack ice like the polar seas. Once sunny, wild prairie, now in summer the land is host to a novel species of grass: maize.

3 1000 km

One inanimate creature of the marine air that can reach regional size is the dramatic hurricane. Here *3* is one example; born in Atlantic or Gulf, it passed over Mexico, to spin out its counterclockwise life in these Pacific waters. Clouds paint the spinning winds, and that calm clear eye looks up at the camera.

By day the marks of mankind are hard to find at this scale. But as earth turns the land into night's shadow, our presence strangely enough becomes visible *2* to the sensitive satellite camera. City lights mark out the eastern half of the United States.

2 1000 km

1 thousand kilometers **1 million meters**

This region, viewed from a low orbit, holds the whole of Lake Michigan; the broad sheet of water, like the flat silted lands around it, was formed by continental glaciers in the most recent geological past, a few tens of thousands of years ago. The day's weather is marked by clouds arrayed in streets and clumps. Though we are looking at the homes of tens of millions of people, the work of human hands is hardly to be seen.

THE LOOK OF THE COUNTRY

This is the scale of countryside, more comprehensive than any single peak or river, yet with a kind of unity. It is the diversity of the earth which is here suggested.

The satellite photos shown here from an ongoing worldwide survey by Landsat display that diversity. That satellite records an area about 185 kilometers on an edge from 500 nautical miles up; the odd shape of the image is made by the earth's turning under the moving satellite. The path of Landsat around the earth is like that of a machine-wound ball of string, westward-tending at each pass so that it sees an endless morning; eighteen days go by before it comes back over any single place once again. Since only two of the colors recorded by the TV camera are visible to our eyes, these pictures do not show natural color. The colors here are coded: Green vegetation is shown in deep red and the gray cities are shown in blue.

The Landsat image at *1* shows a varied countryside in New Zealand. Quiet rivers flow down from the snow-capped peaks of the Southern Alps. The city of Christchurch is a blue patch where the smooth coastline breaks off into hills. At *2* all is empty and dry: the Baluchistan Desert crossed by the border between Iran and Pakistan. In *3* the island bead of Hawaii, pierced by great volcanos, adorns the Pacific. Another much larger crater can be seen at *4* in eastern Canada. Not volcano-formed, it is the scar of an impact made on this rocky surface two hundred million years ago by a little asteroid about three kilometers across. Now it is used as a reservoir.

1

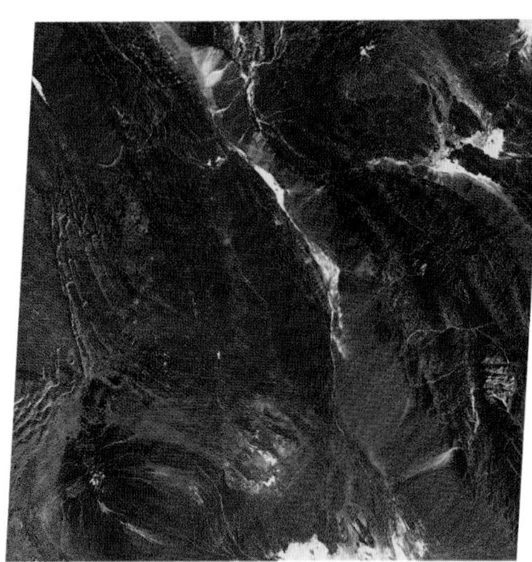

2

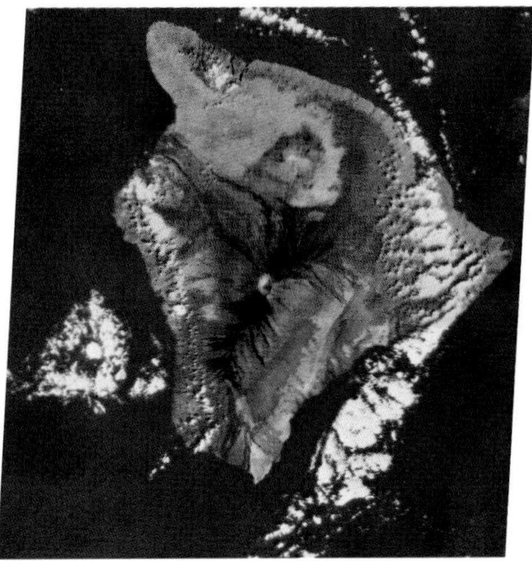

3

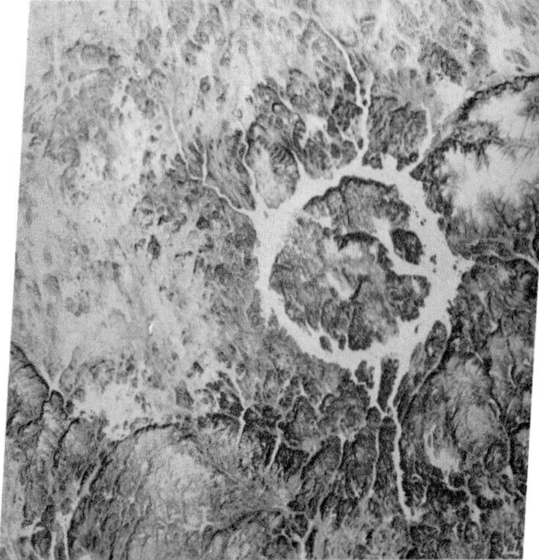

4

100 kilometers

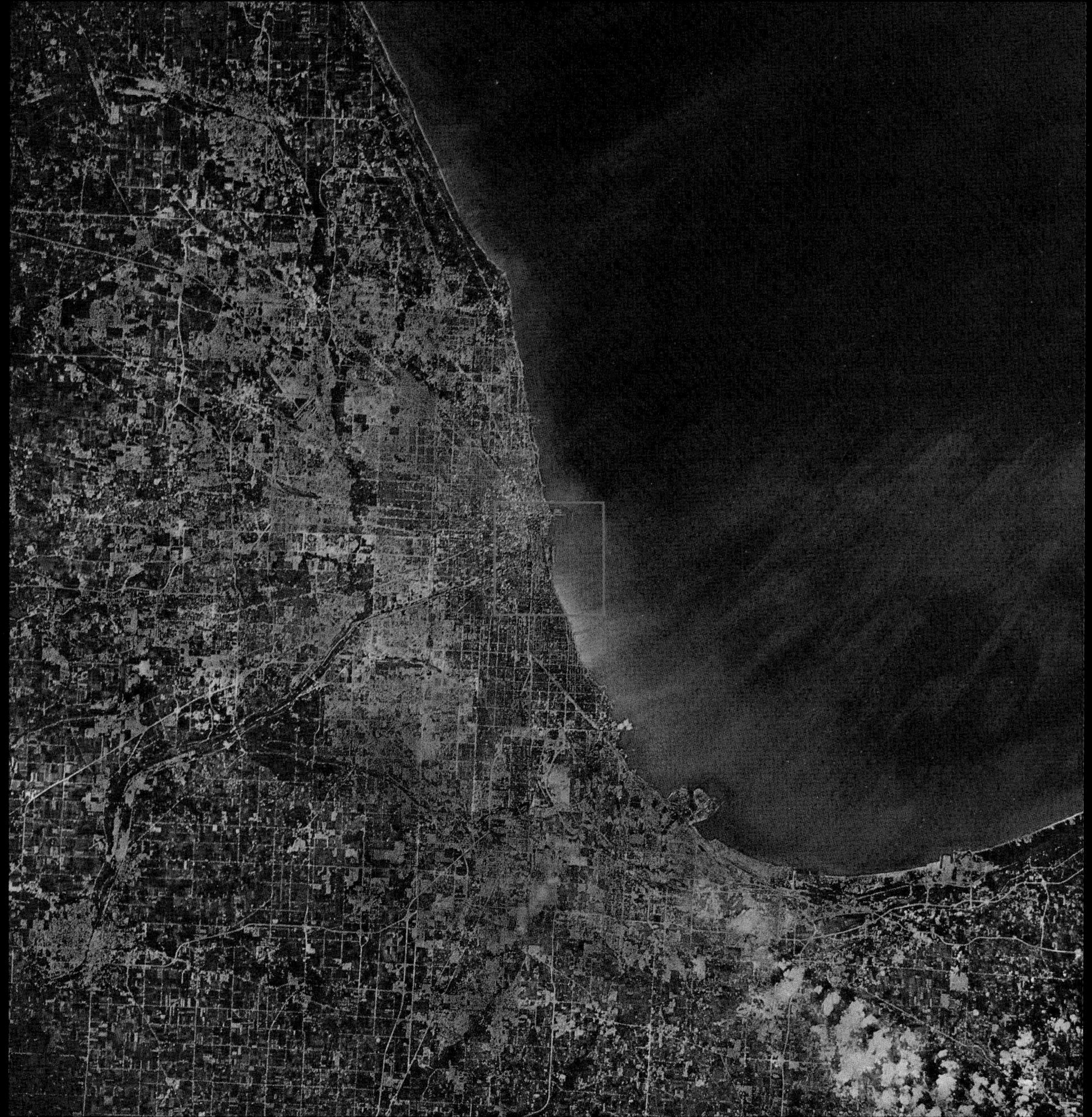

The metropolitan area of Chicago nestles at the south end of the lake. On a day like this, someone walking along the street might have looked up to a blue sky; but the camera plane was flying so high it would have been hard to pick out. The lattice visible among so many blurred streets is the mile-square grid of wide Chicago boulevards.

At this scale the collective work of our species is easily admired. In the oblique aerial view *1* is seen the strong pattern of wheat fields west of McPherson, Kansas. The roads, woodlands, and houses are only affixes to these broad acres, where sunlight is caught for bread. Every city implies a scene something like this.

1

2

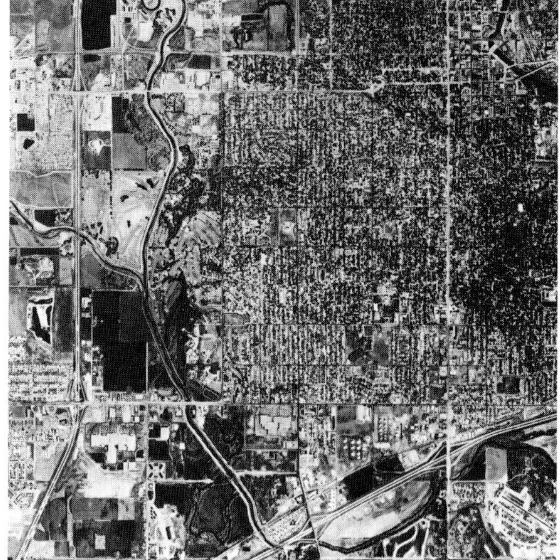

3

Here the essential fields enfold a small town *2*, whose striking shape proclaims its planned origins. It is the walled military city of Palmanova, built quickly around 1600 in the approved style of the day for fortresses. It was built a hundred kilometers east of the Grand Canal to defend the frontier of the Republic of Venice.

Sioux Falls, South Dakota, is another geometrical city among wheat fields *3*, though very different.

Ten kilometers horizontally across the level earth is a modest, even a homey distance for dwellers in city or country. But vertically it presses upon the limits of the earth. Mt. Everest reaches only that high above the distant sea; the deepest trenches of the Pacific depths plumb only ten or eleven kilometers beneath the ocean surface. Correctly scaled profiles of Everest and of the deep Tonga Trench are shown in *4*.

4

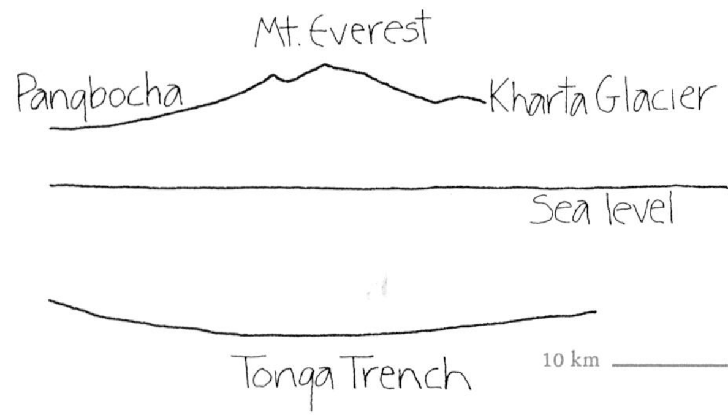

10 kilometers **~6 miles**

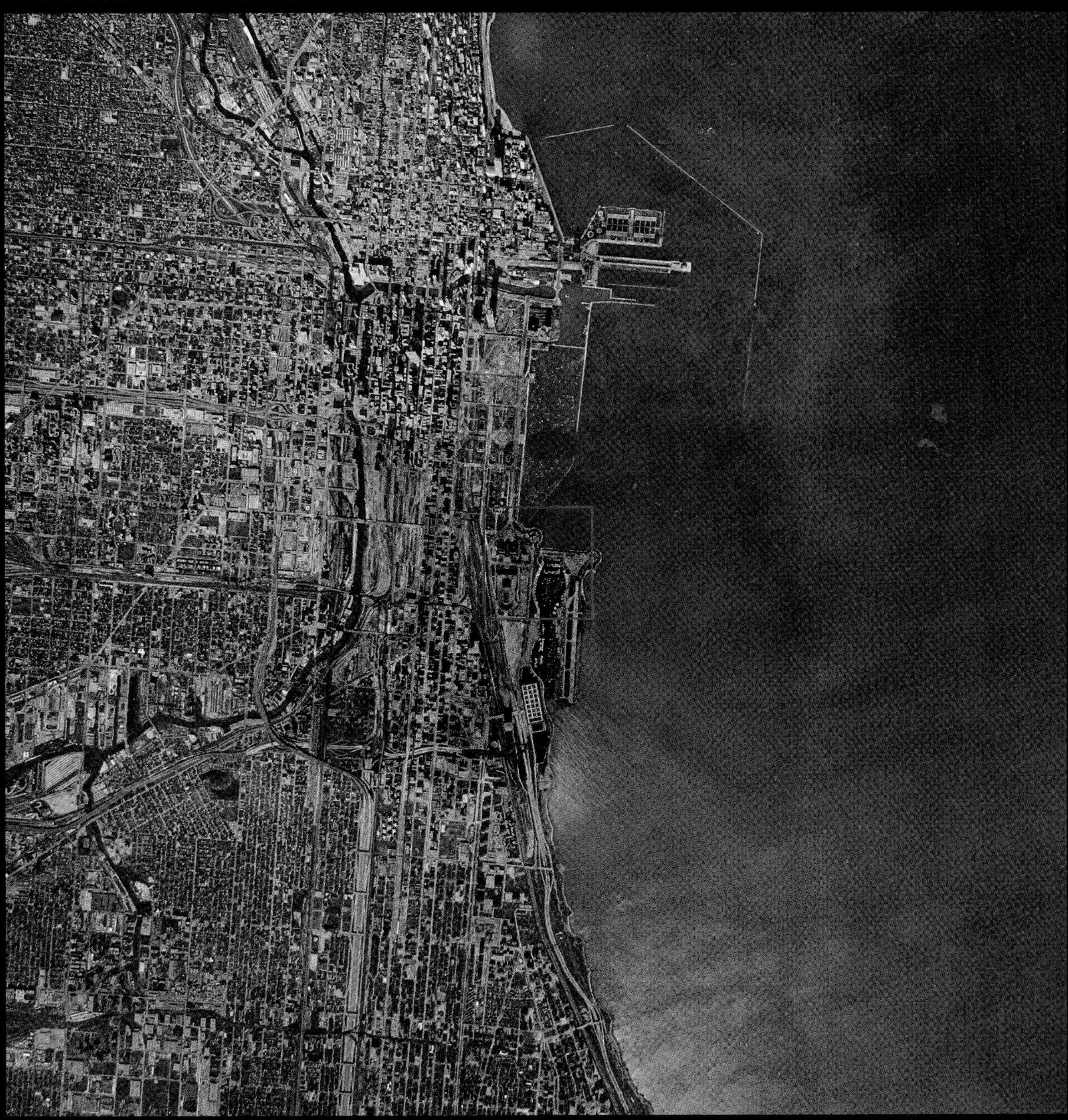

The heart of the city appears, place of home and work for a million people. The whole structure shown here—city districts, parks, harbor—is familiar to them. The conflagration of 1871 burned the city of wooden houses which then lay within this square. Most of the detail shown is newer, though the street and railroad layout survived the fire, as in the future they will outlive most of the individual buildings.

Just north of the stadium-horseshoe on the opposite page is seen a large black-roofed building, rather in the form of an H. It is the Field Museum, this neighborhood's extraordinary treasure house. In the photo *1* we see one of its holdings: an oval of tanned buckskin, painted with mineral pigment. It was acquired by the Museum in 1906 from its owners, the Skidi band of the Pawnee people. Once the Pawnee lived along the Platte River in Nebraska; in 1875 they were removed to Oklahoma. The skin is a star chart, on which recognizable patterns have been drawn, less in direct mapping than as a symbolic display of important forms. It had long been used as a ceremonial object within the memories of our informants, part of a medicine bundle. It is not hard to make out the Milky Way, the Pleiades, the constellations Taurus and Corona Borealis, and more. "*The Skidi were organized by the stars*," said Running Scout; there is evidence that their seventeen villages were once carefully set across the grasslands in the pattern of certain stars, the Pawnee Constellation. Time has eaten away at their unwritten lore; but that they, too, noted with care and delight something of the celestial order, and sought a visual model, is witnessed by the leather chart.

2

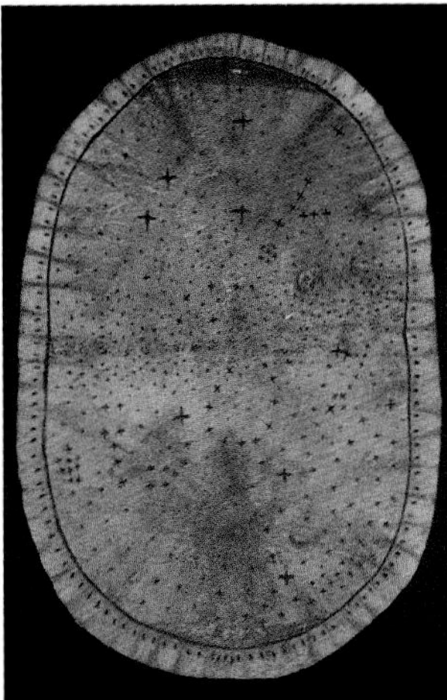

1

3

About a hundred thousand people are gathered to watch a football game at Soldiers' Field.

Leave this Chicago neighborhood for New York City and the Hudson, west of Manhattan. There an eloquent stretch of spun steel cable tensely spans the river. Would you form a close estimate of one kilometer? That is the clear horizontal span of the George Washington Bridge *4*, opened fifty years ago. The longer linear forms we have made—highroads, pipelines, and the Great Wall itself—reach other scales, up to a few thousand kilometers, but they all lack the structural integrity of a great bridge. It soars, while they lie upon solid ground, one-dimensional and repetitive. No structure yet built stands so tall as one kilometer.

4

Just to the right of the kilometer square, at the very edge of the neighborhood, is the Adler Planetarium, with its remarkable collection of astronomical instruments. Among them is the beautiful contrivance *2* in ivory and brass. It is a geared model of the planets of the solar system, made in Benjamin Martin's London shop about 1780. Such a device can hardly be made to consistent scale, but the planets execute their proper revolutions, driven by wheelwork within the brass drum. With no Uranus beyond its Saturn, it was surely completed before the much-heralded discovery of that telescopic seventh planet by William Herschel in 1781. In many a drawing room of the Enlightenment such a device was a prized ornament and a visual model of the world order of Newtonian philosophy. The planets are set here in the relative positions they occupied on the October day of the journey in this book.

1 kilometer **1 thousand meters**

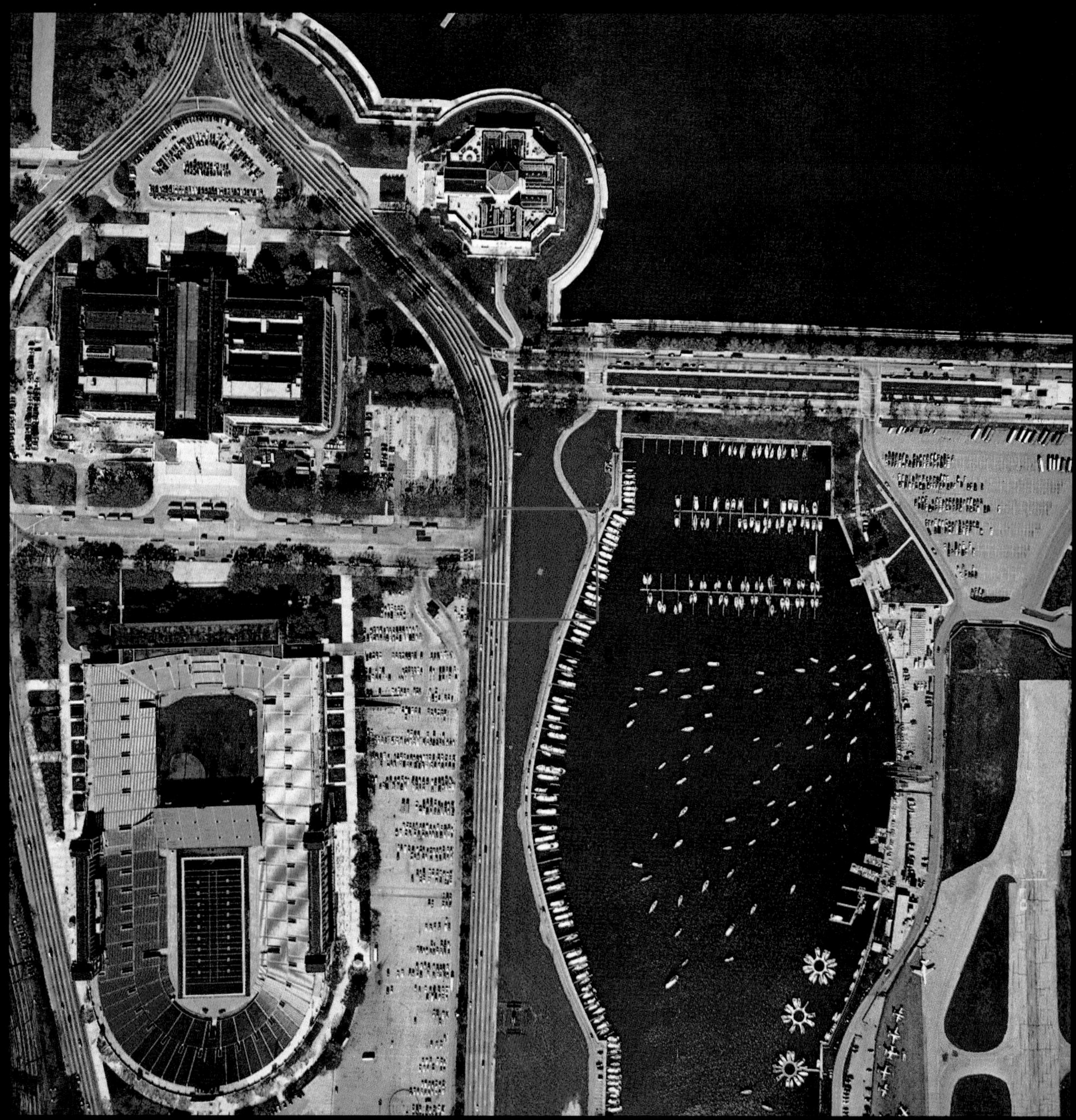

Now we look at a view that is not a maplike tracery of symbols, but a scene of familiar places within the city: Lake Shore Drive, Soldiers' Field, an airstrip, boat docks, museums.

10^2 meters

A PARK

1

3

2

The class of 100-meter objects includes single human artifacts; at larger scales the artifacts of our kind are complexes, like cities and fields of crops. Living organisms too—again, not their communities, not forests and coral reefs, but single individuals—reach this scale, but they have never gone beyond it.

This long-lived giant sequoia *1* is among the largest of living forms, around 80 meters high; a few less robust trees cross 100 meters. No individual forms of life are larger than the great trees; there may be longer ground vines or giant seaweeds, and certain groves of bamboo with some claim to being one organism might extend farther. But the 100-meter scale is a sound rough limit for the size of living beings.

Two more vertical structures of the 100-meter scale are shown here: The Saturn V moon rocket *2* is 110 meters from nose to bottom of jets, while the Statue of Liberty *3* stands 93 meters torch-tip to ground.

Many buildings do rise higher than 100 meters. That famous Paris tower in iron (built by the same man who designed Liberty's skeleton) reached its 300-meter height in 1889. The tallest building in the world rises 450 meters, just 3 kilometers away from our picnic; a few thin masts, towers, and chimneys—metal analogues to the vines and the kelps—now stand even taller than the 450 meters, though none of them goes much beyond 600 meters height. Buildings can easily be 100 meters wide. Examples are the Taj Mahal *4*, whose four minarets form a 100-meter square, and the superb limestone theater bowl at Epidaurus in Greece *5*, which is just under 120 meters in diameter.

Ships are the largest mobile artifacts. Even a good 30-meter sailing vessel can go anywhere at sea; Captain Cook's *Endeavor* was just that length. But the clippers were 60 meters and more, and a big full-rigged tall ship of today, the Coast Guard training vessel *Eagle* in *6*, is overall 93 meters long. The biggest liners and warships are around 300 meters, and the longest ultra-large tankers stretch a little past 400 meters from stem to stern.

5

6

100 m

4

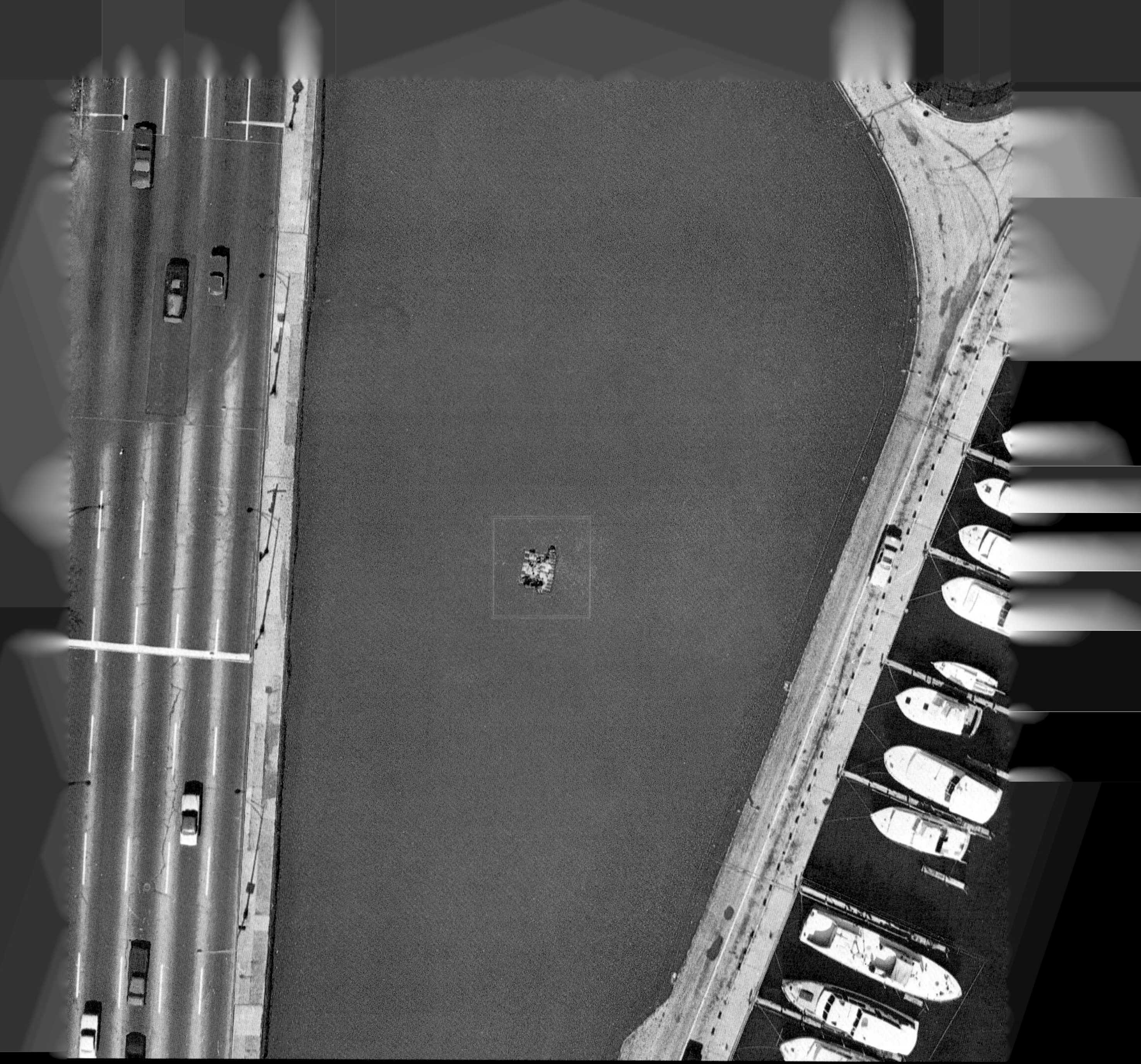

The picnic in the park is not far from the roaring highway and the boats at their docks The picnickers can enjoy a sense of privacy all the same, for no one else is near. Were people evenly spread over all the world's land area, these two could lay claim to six times the area of this whole square. To raise their own grain, they would need to cultivate only this

1

10 m

The biggest animals that have ever lived are the blue whales, a species threatened in our time. Big whales in general go beyond this 10-meter length, up to 30 meters for the largest forms, though no animal life has ever approached the 100-meter scale. The beast of *1*, a sauropod dinosaur known as *Diplodocus carnegii*, was about as long as a land animal has ever grown, just over 25 meters, fully whale-length, though its whiplash tail cannot compare with the ponderous flukes of the sea-supported mammal. Its time was about sixty-five million years ago, its place the wet lowlands that are now the arid Dinosaur National Monument, in Utah. The largest creatures of our own mammalian class, less alien to us than any scaly reptiles, were giant hornless rhinos of twenty-five million years ago. Treetop browsers across the Asian savannah, this particular species is reconstructed from fossil finds in the Gobi desert. *Baluchitherium 2* could have looked comfortably right over a procession of circus elephants. Its length is 10 meters, its shoulder height 5.5 meters.

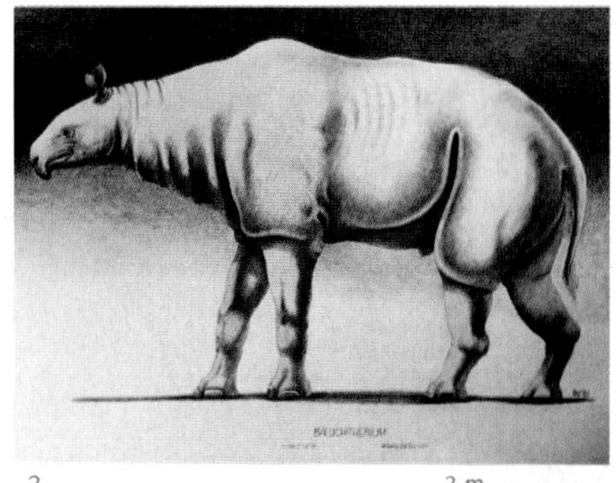

2

2 m

To the nearest order of magnitude, a dwelling house is 10 meters wide and 10 meters high. This pleasing artifact in *3* is about like that: It is the frame house of John Wilson, a country printer, built in Deerfield, Massachusetts, about the year 1816. In the same way, an airplane is a 10-meter artifact. That includes light planes from the original Wright Flyer to this much-used modern metal-skinned aircraft *4*. Even the powerful jet behemoths have not yet come very near 100 meters in any dimension. The smallest aircraft, like the largest birds, are 3-meter fliers. Once there was a 10-meter flying reptile; it must have been a glider.

3

2 m

4

A man and a woman are at a picnic in the park. This picnic is the center of every picture

10^0 meters

This is the scale we know best, our own.

2

1

3

4

5

6

7

This is the scale of human companionship, conversation, touch: A man is asleep on a warm October day. Around him are necessities and pleasures for mind and body. Between this image and the next frame inward, the size of the image would for once match the size of what it represents. "Of all things man is the measure," wrote Protagoras the Sophist

10^{-1} meters

HAND SAMPLES

Hands have grasped many different tools. The natural world is even more diverse at a handy scale. Some of these images continue the line of human artifacts, some that of living forms. Here are tools *1* kept at hand by the scholar-saint Jerome, who rendered the Hebrew Scripture into the Latin Vulgate Bible, as they were imagined a thousand years after his time by the Florentine painter Domenico Ghirlandaio in the fifteenth century. But the calipers in *2* are real, made of brass by Paul Revere of Boston to check the diameter of cannon balls. The large watch *3* is the first successful chronometer, a masterpiece of complex hand craftsmanship and design by John Harrison, who with it won the prize long offered by the British Admiralty for a means to measure longitude from shipboard. Hand size is also the size of many diverse living forms. These blossoms evolved to attract their insect pollinators by day. The evidence is at hand, for by the ordinary light useful to human vision *4*, the flower is an unmarked plain yellow. But photographed in ultraviolet light to present what the insect sees *5*, the blossom is tellingly marked with a dark pattern. The bullfrog in the pond *6* is about ten centimeters long. That shrimplike animal with big luminous eyes 7 is called krill. It is the staple food of the plankton-sieving blue whale of the Antarctic Ocean.The whales are now so few that the krill abound; it may be that the total living mass of this species outweighs any other animal species, even our own. The shrew in *8*, fierce incessant hunter of insects and worms, is about the smallest of the mammals, at the other end of the sequence from the blue whale. It must eat steadily or die; there is no respite from swift loss of body heat at its small size and great relative surface.

1

3 5 cm

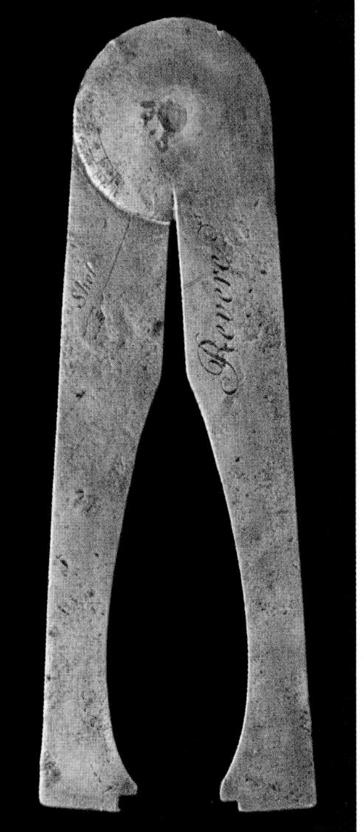

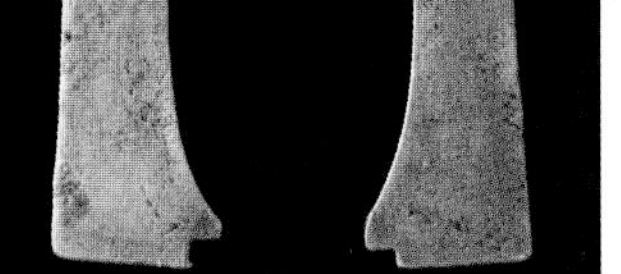

2

4

5

6

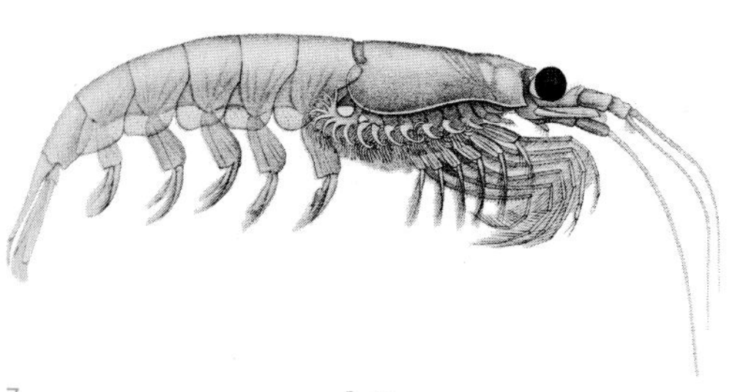

7 5 cm

8

0.1 meter **10 centimeters**

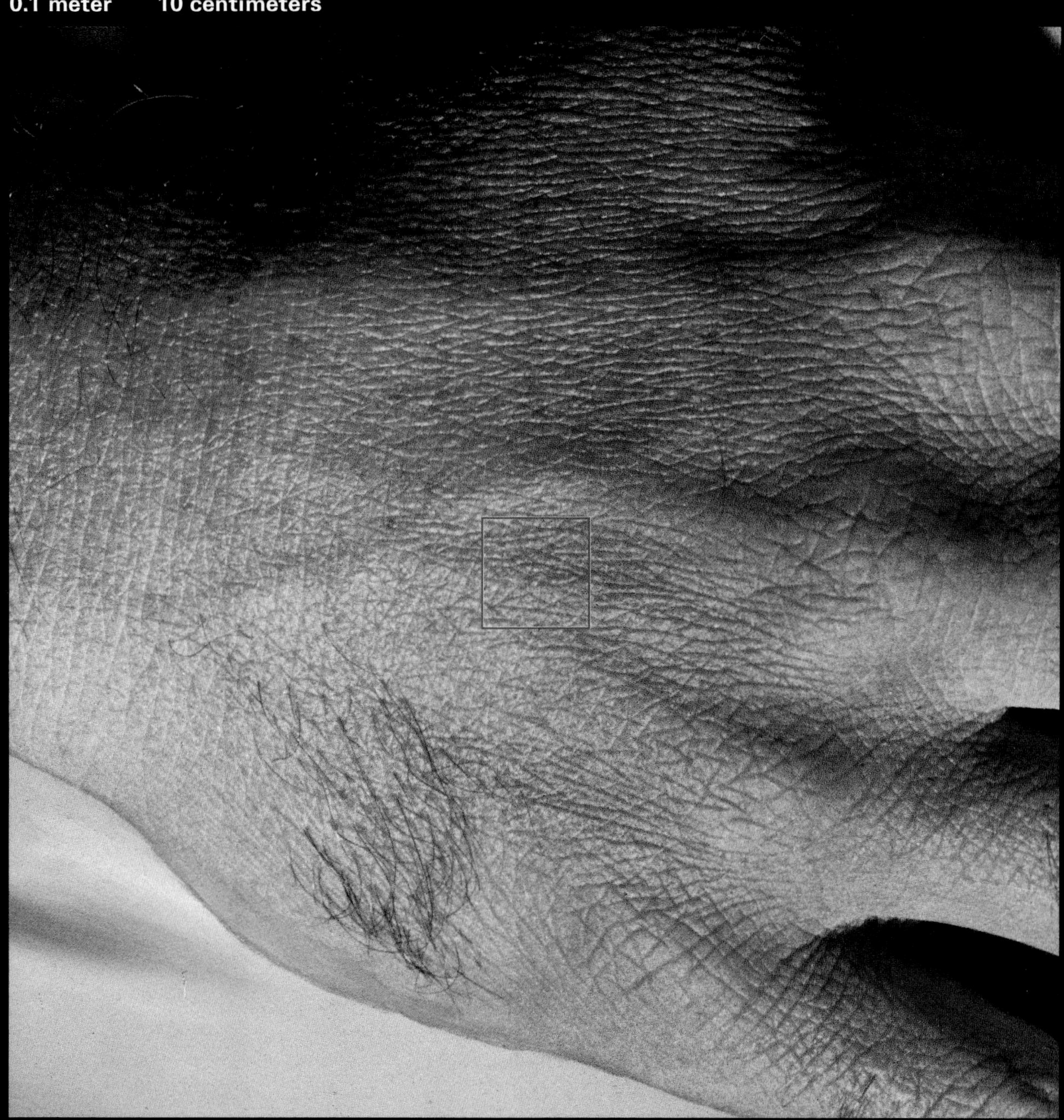

The scale is now intimate: This is the look of the back of your own hand, a little enlarged. That animate structure, guided by eye and mind, joined over time by many another in the human endeavor, has fashioned all the representations we have of the world, including this of the hand itself.

1

2

Nimble fingers play over keyboards everywhere: The typewriter *1* provides Japanese syllabary characters, *katakana*, until the shift key is depressed. These handsome cast silver buttons *2* belonged to John Adams of Quincy. Living things are represented by a small mushroom *3* from the wet woodlands, a bombardier beetle *4* directing its explosive defensive spray accurately against the intrusive forceps pinching a left foreleg, and a small hydromedusan *5*, almost transparent and swimming freely.

3

4 1 cm

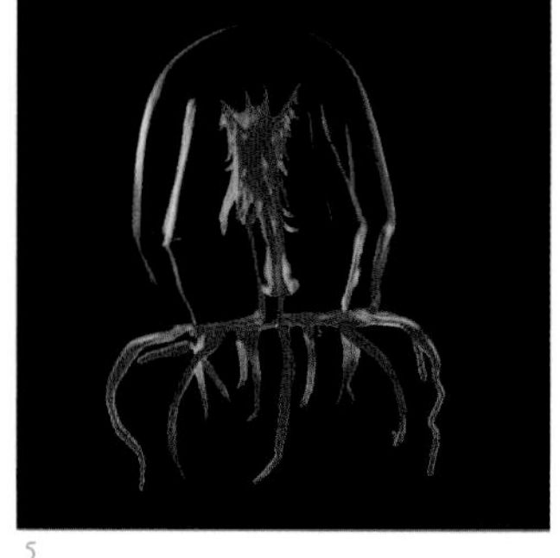

5

A fingertip has no pattern of crinkled folds, but it bears the familiar set of whorled ridges that form the fingerprint *6*. A network of lines marks the back of the hand, only one of many polygonal networks found in nature, all obedient to powerful topological constraints. This network *7* is a foam held between two glass plates, seen at almost natural size. In it, three bubble surfaces meet at each corner in the mechanical balance required by the forces of surface tension. The interplay between overall constraints and local balance results in an abundance of pentagonal faces. The foam pentagons are surely related to the deep cause of the fivefoldedness seen often in life, for instance in the five fingers of the hand.

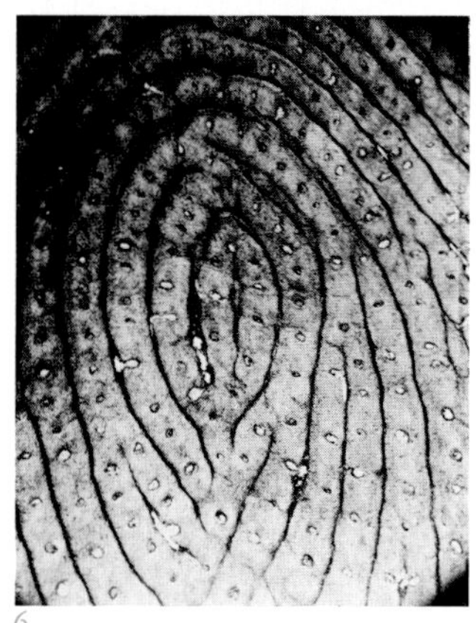

6

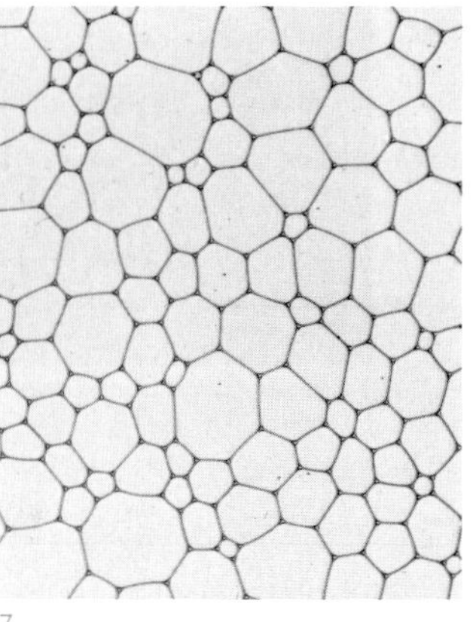

7

1 centimeter

A searching look at the skin as if through a strong magnifier. The creasing is both the sign and the means of the skin's flexibility.

These microscopic scenes are so near the edge of ordinary vision that they are recognizable even in novelty, hidden surprises within the commonplace. Robert Hooke drew the piece of handkerchief linen shown in *1*, and remarked: "*A piece of the finest Lawn I was able to get . . . and yet through an ordinary Microscope you may perceive what a goodly piece of coarse Matting it is.*" "*The small seeds of Poppy,*" he wrote, "*for their smallness, multiplicity, and prettiness . . . deserve to be taken notice of.*" After three centuries, they still come from the bakery just as Hooke described them *2*: "*. . . being not above 1/32 part of an Inch in Diameter . . . curiously Honey-comb'd all over.*" Microscopy is much more versatile now; in *3* is a living transparent pond-water organism, *Stentor*, fringed by cilia, seen in a modern light microscope, and in *4* a parasitic mite, tiny on the neck of its termite host, viewed in depth by a scanning electron beam. Table salt, shown in *5* also through the scanning electron microscope, is unmistakably mineral, as faceted and crystalline as the one-meter topaz.

The piece of crude hardware *6* is in fact one of the delicate adjusting screws that rim the balance wheel of a fine watch. In *7* next to the wing of a gnat is the smallest motor ever made. Fed alternating current through the fine copper leads, this sub-pinhead of an electric motor spins dizzily. It is about the size of the head of the screw in *6*. But in today's most intricate artifacts, no parts move save electrons. This is a small thin square of silicon *8* that bears about a hundred thousand functional circuit devices. It includes the memory of a powerful computer.

THE EDGE OF VISION

. . . a sincere Hand, and a faithful Eye, to examine, and to record, the things themselves as they appear.

Robert Hooke,
Micrographia, 1665

1 0.5 mm

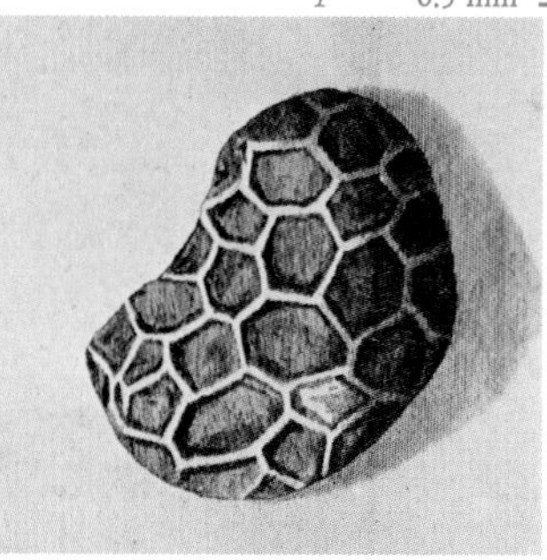

2 0.5 mm

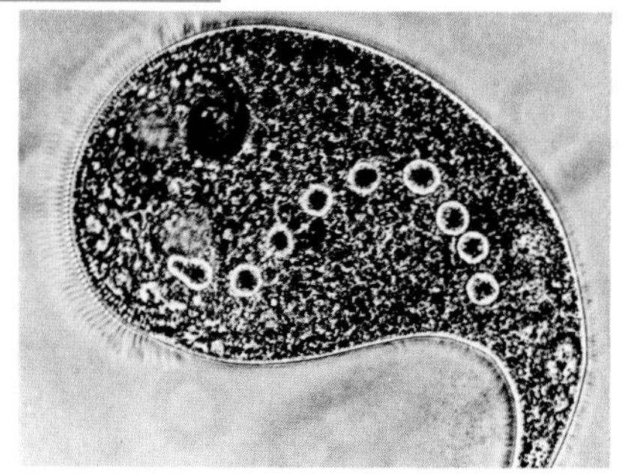

3 0.5 mm

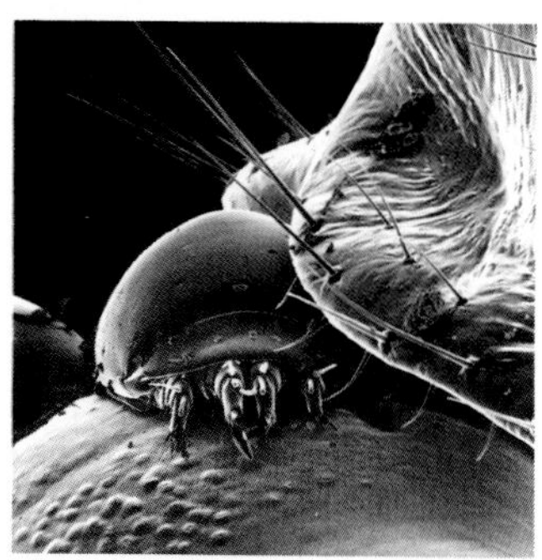

4 0.1 mm

5 0.5 mm

6 0.5 mm

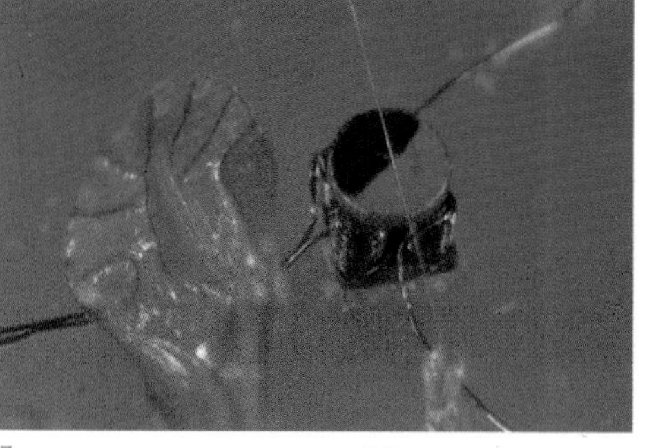

7 0.5 mm

8 5 mm

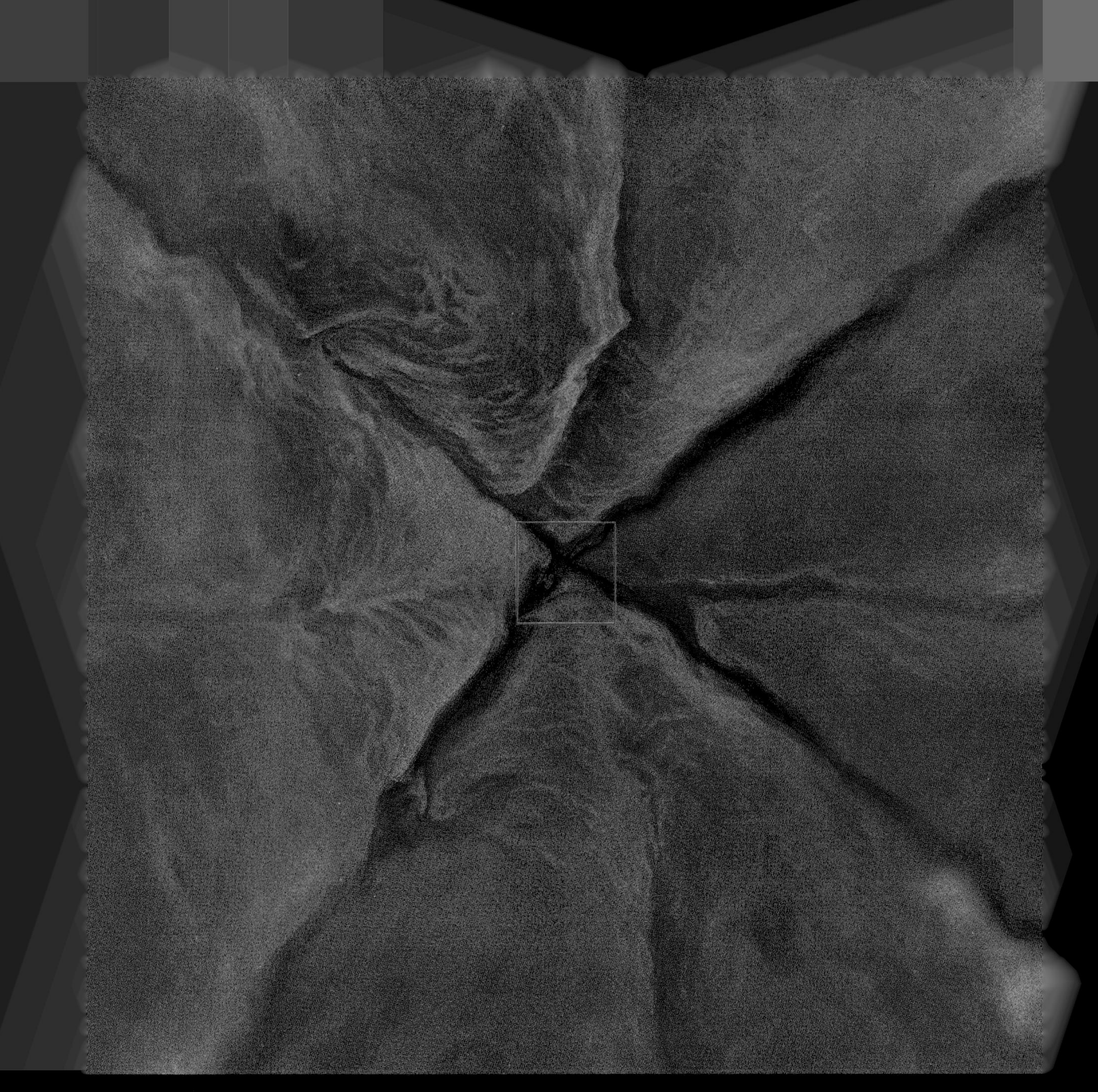

Here we share the world of the microscopist, who has unlocked so much of nature. For each image still closer in than this one, we come nine-tenths of the remaining distance toward the inner end of our journey, just below the skin of the man, within a cell passing along

10^{-4} meters

The thin layers of the skin that lie above the capillary blood vessels are shown in magnified section in *1.* In some places, cell layers are separated; the outermost cells are in the process of shedding. Hair shafts penetrate these layers.

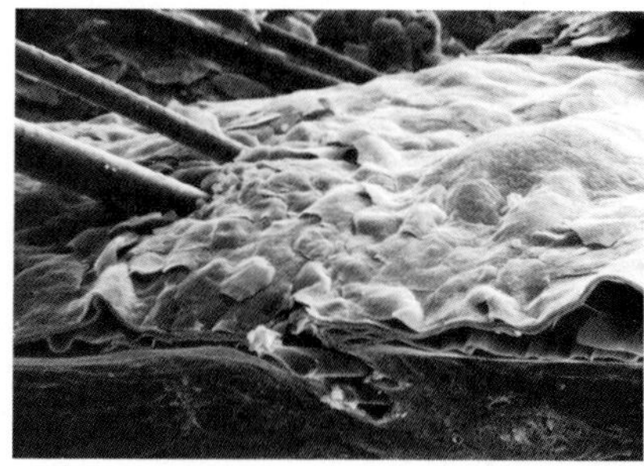

1 100 μ

2 200 μ

In *2* is a well-known artifact of delicate detail, one we know better by ear than by eye: the grooves of a long-playing record. The very small portion seen takes only a couple of milliseconds to play. The wavy form of some grooves shows the scale of the recorded sound. The frequency range is a few thousands of reversals per second.

Another living freshwater ciliated protozoan is shown in *5*, replete with newly ingested strings of blue-green beads. Those are filaments of widely found photosynthesizing bacteria.

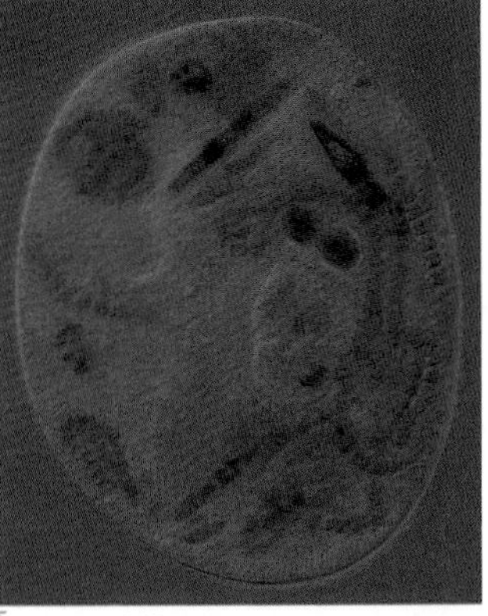

5 100 μ

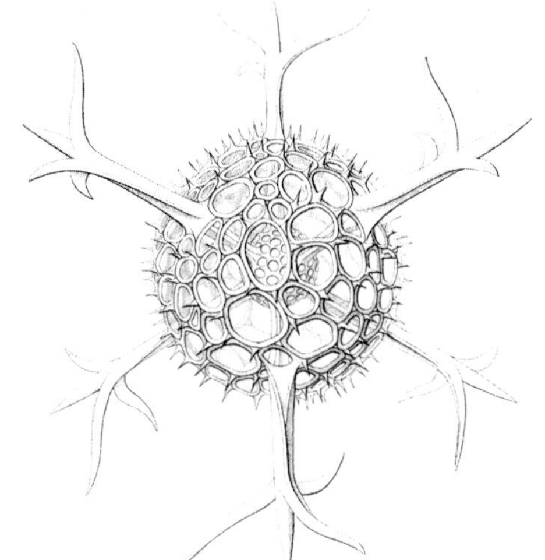

3

4 200 μ

Radiolaria are gorgeous minute marine creatures kin to the amoeba and important members of the zooplankton, especially in tropical seas. These pieces of exquisite lacework are the tests, or skeletal frameworks, of some forms. Unlike calcium-based bone and shell, these are made of translucent silica. The sea bottom contains such jewels as a major constituent over large areas. The photo *4* shows an example taken recently from the bottom ooze near the Azores; the drawing *3* is chosen from the compendium published by the scientist, artist, and philosopher Ernst Haeckel as part of the report of the renowned expedition of HMS *Challenger*, which sounded the ocean depths a century ago.

0.1 millimeter **100 microns**

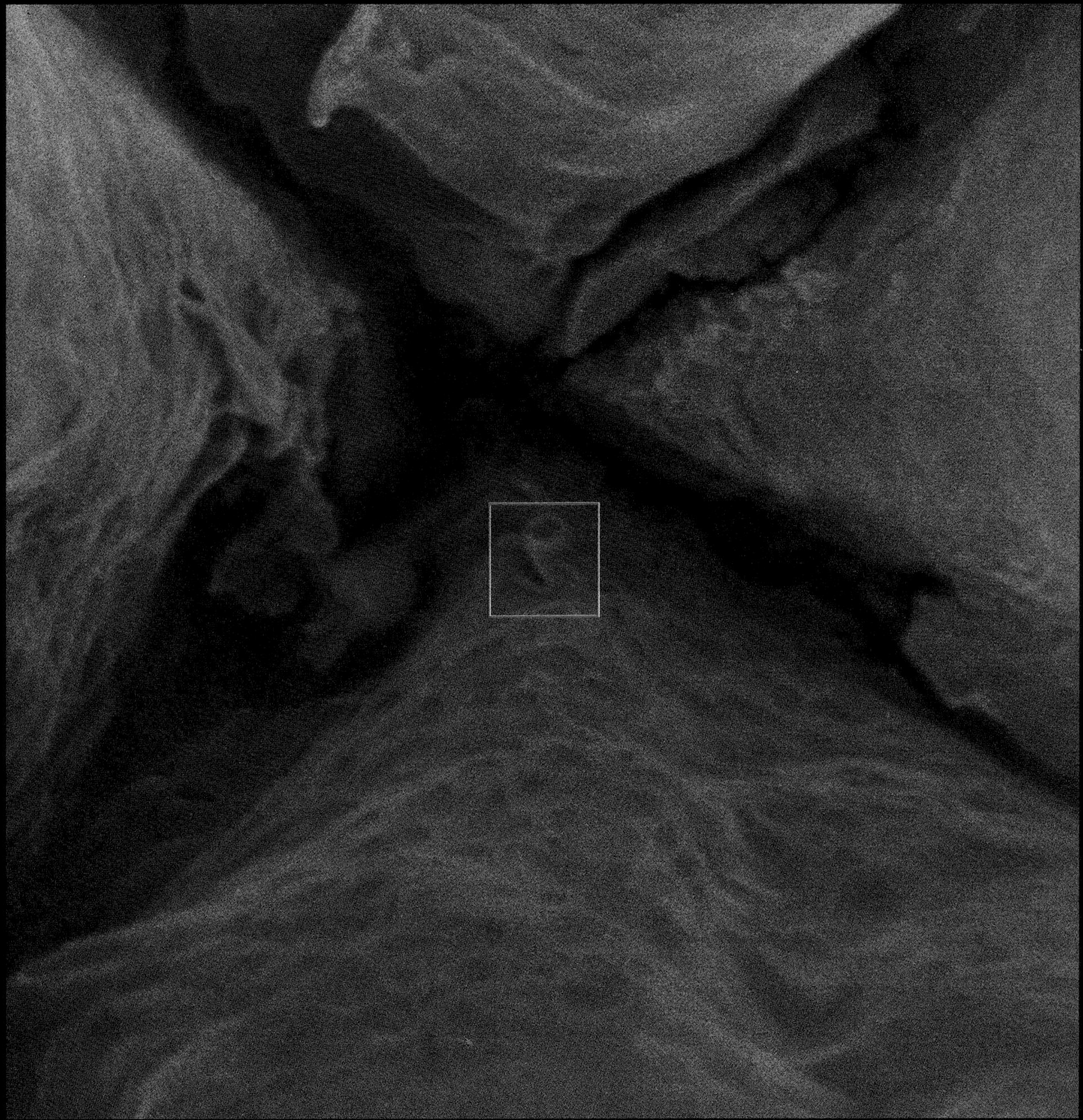

Unexpected detail appears; we can scarcely orient ourselves. Deeper still, we enter an intimate world within, as unfamiliar to us as the distant stars.

STRUCTURE AND LIGHT

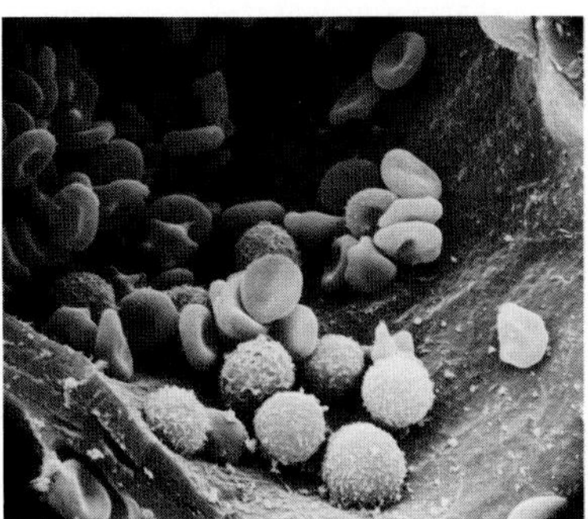

1 10 μ

A small blood vessel is opened up to the probe of the scanning electron microscope *1*, exposing a strew of blood corpuscles, both the disks of the oxygen-bearing red cells and the ruffled spheres of the white.

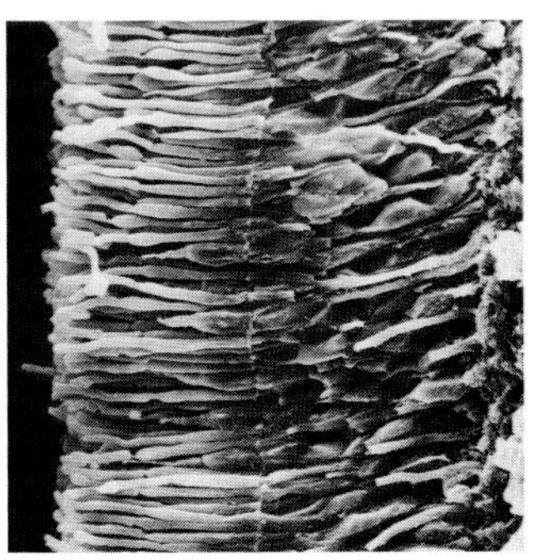

2 20 μ

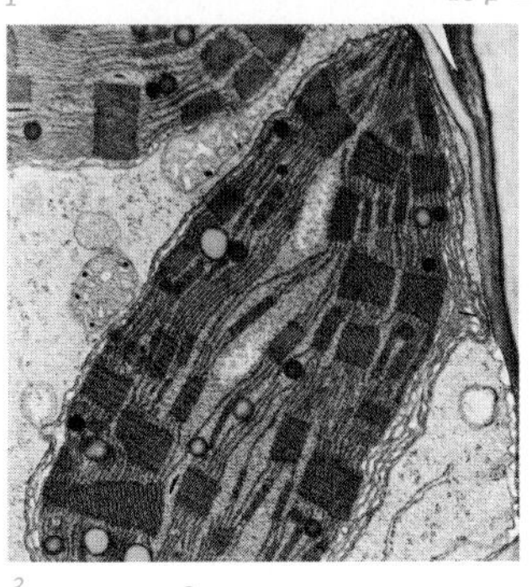

3 2 μ

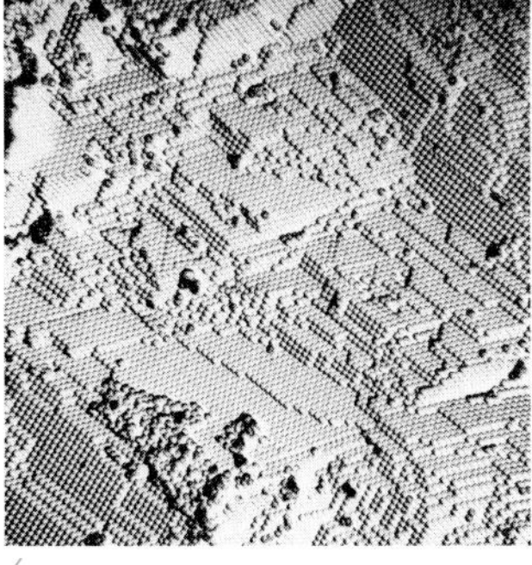

4 10 μ

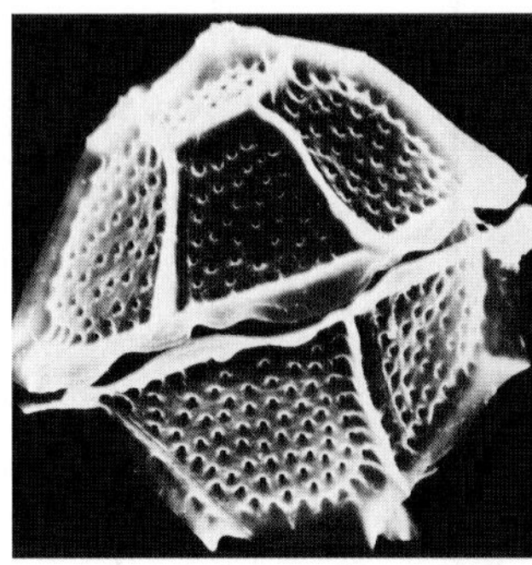

5 10 μ

6 10 μ

This is a decisive scale, at which the cells that form tissue are fully distinct. This scanning electron microscope photo *2* shows a section of the retina of the eye of a pig, very like the same structure in humans. Light comes in from the right, to travel to the photosensitive layer along the left edge of the picture. From this detailed look at a most remarkable instrument it is evident that the individual retinal cells, like the distinct grains on a photographic plate, must set some limit to our visual grasp of fine detail. It is striking that the retinal limit, on the order of a micron or two, is only a little above the intrinsic limit fixed by the nature of visible light—its wavelength of around half a micron. Vision rests on the interaction of the specialized cells of the eye with light. But life itself depends on the photosynthetic energy won by green leaves through the absorption of light in special organelles, called chloroplasts. A chloroplast in a corn leaf cell is shown in *3* in thin section under the transmission electron microscope. The green pigment chlorophyll, which traps light energy to begin a long chain of chemical processes, is held in the stacks of disks seen here in cross section. Each disk is a flattened enclosure of pigment and protein.

Fine structure is present at micron scale in the mineral world, too. This scanning electron micrograph *4* shows a portion of precious opal. Opal is plainly a regular array, made of tiny uniform balls of silica. Their spacing is 0.2 micron, which enables them to interact in a repetitive and reinforcing way with visible light not far from that wavelength. The result is the fire of the opal; the light is sent in varying directions according to color. The same phenomenon is scaled down to atomic size in ordinary crystals of matter, all of which are such regular arrays on an atomic scale. Their atomic spacing is not 0.2 micron but a thousand times smaller. Thus all crystals are in effect opals, but opals only for the proportionately shorter waves of x rays, not for visible light itself. That is the basis for the powerful x-ray analysis of crystalline matter.

These are two more examples of condensed complexity. One is a strange freshwater organism *5* with an intricate armor of cellulose. The other is a modern artifact that just begins to rival the biological level of minute pattern. It is an experimental computer device *6*, its carefully formed walls of pure gold.

10^{-5} meters

10 microns

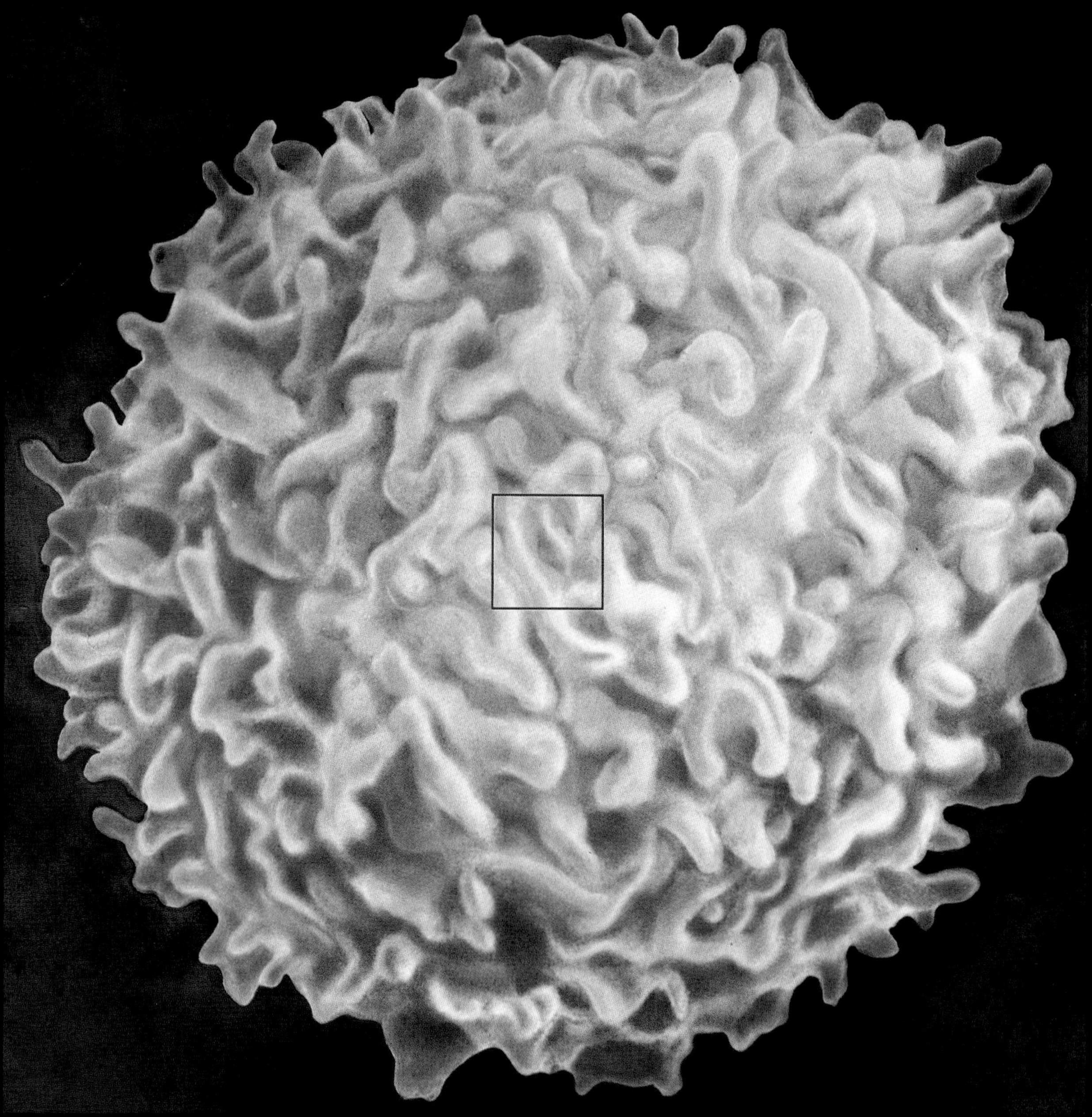

We pass through the living skin to enter a capillary vessel, where blood oozes by. Most blood cells are the small, incomplete, short-lived disks that give red blood its color; this white cell, a lymphocyte, is a long-lived participant in the complex cellular and chemical strategy called the immune system, the body's defense against infection.

Bacteria lack not only nuclei but the other specialized organelles found in larger and more complex cells. Their chlorophyll and oxidative enzymes are not neatly packaged as they are in the cells with true nuclei.

The smallest living cells are only about 0.1 micron in diameter. Few of the simpler cells grow as large as 10 microns. Nucleated cells measure from tens of microns, like the typical cells of the human body and of multicellular animals and plants, up to a millimeter for the free-living single-celled creatures, the protists of soil and sea, crowded with efficient organelles. The egg cells of animals can be even larger.

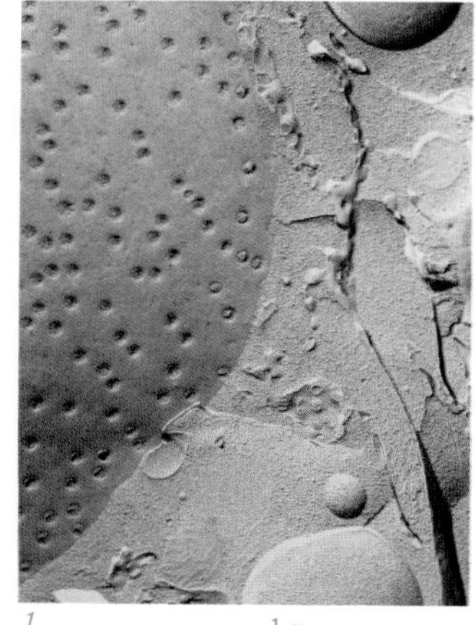

1 1 μ

Nuclear pores are no simple openings but complicated molecular passageways. In the scanning electron micrograph of *1*, the nuclear surface within a cell of an onion root tip holds many pores. Recent techniques of freeze-fracture expose the surface for a clear view.

The human cell holds its genetic legacy safe inside the nucleus. That long message is stored on complex scrolls formed of protein and DNA, the forty-six chromosomes. Their message must be consulted at length between cell divisions, duplicated, and carefully partitioned upon division as legacy to the two new cells. In the electron micrograph of *2*, one human chromosome is seen doubled, fairly early in the ballet of cell reproduction. The impression here is that the long molecular message is still partly unrolled.

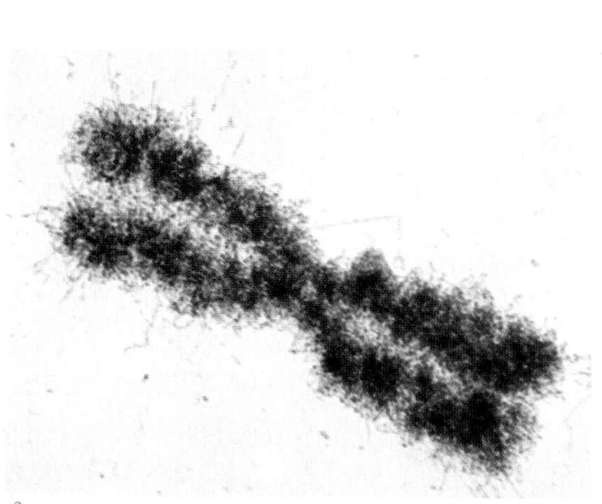

2 1 μ

3 0.3 μ

Every plant cell and every animal cell has a nucleus. But the old families of organic evolution, the bacteria and their kin, have none. The electron micrograph of *3* shows a section of one common bacterium, an active and ubiquitous form in soil and freshwater. Its central region is least dense; within it the DNA, the small cell's essential instructions for self-duplication, is present as one single strong unmounted loop of molecular tape.

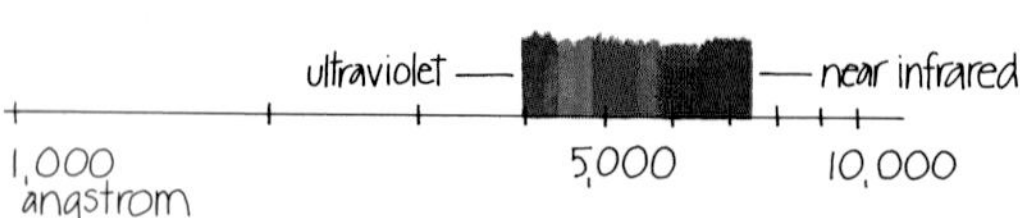

4

The wavelength of visible light is somewhat smaller than living cells. From our own vision, and from the green of plants, we know how much the interaction of cells with visible light means for life on earth. The wavelength scale is drawn in *4*, with color marked along the scale. Bright light at each wavelength will stimulate the color sensation marked, though seeing the color does not always imply the physical presence of light with that wavelength. Color perception is much more complicated than a simple wavelength scale.

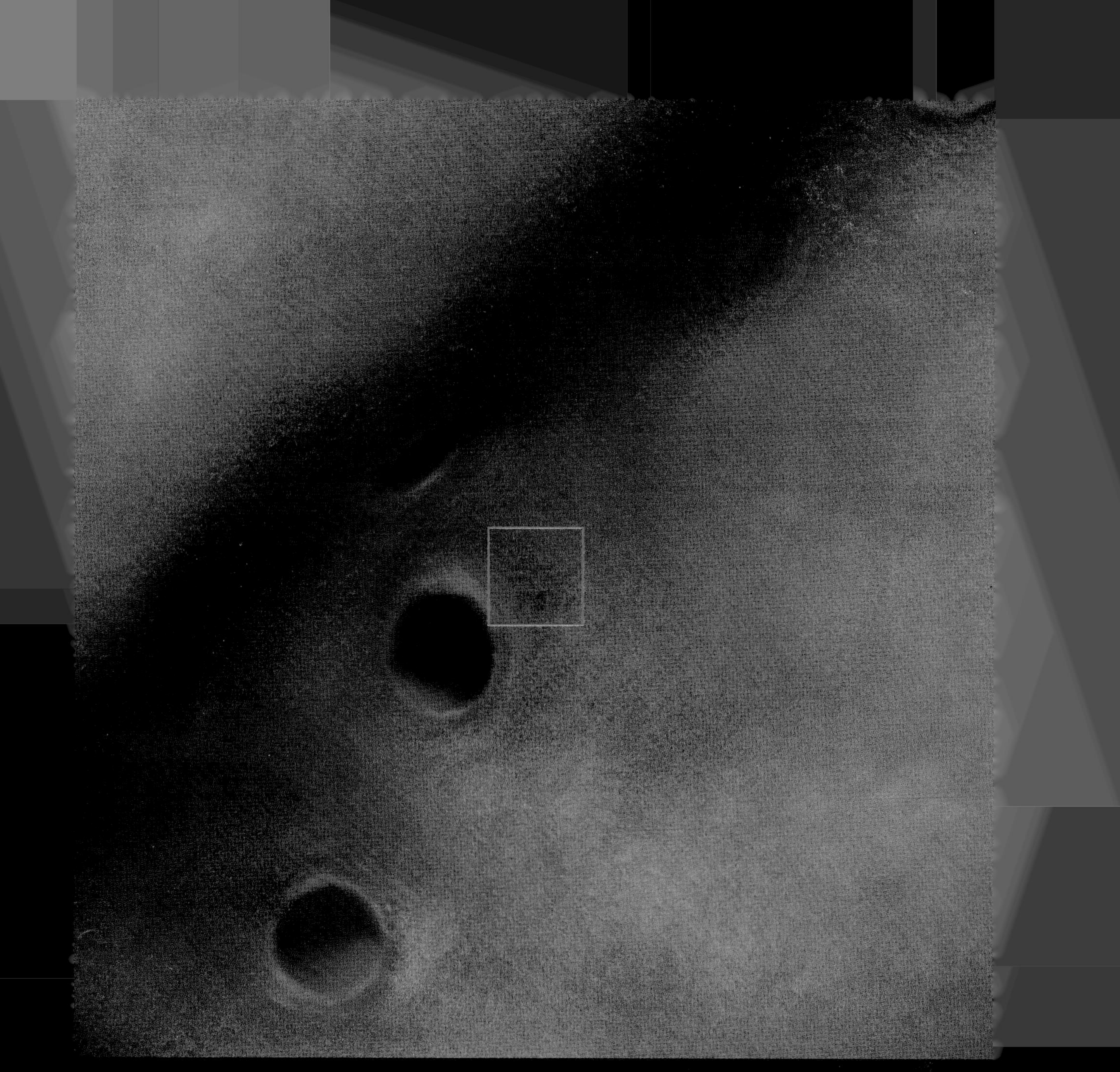

We are inside the ruffly lymphocyte, only to face another surface, a protective membrane within the cell that encloses its nucleus. The minute pores allow materials from within to enter the larger volume of the cell. Every complete cell has such a nucleus, whose molecular products inform the entire life of the cell. In one human body are a hundred times more

THE VITAL MOLECULE

In *1*, a single skein of DNA appears in an electron micrograph. This double helix (two free ends can be found at top and bottom) is a few dozen microns long, perhaps a hundred thousand rungs of the molecular ladder. The center of the image shows the minute container of protein from which the DNA has spewed out.

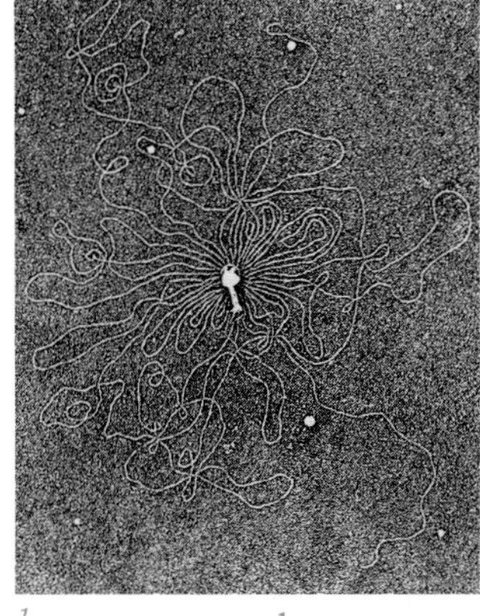

1 1 μ

More steps along the way from a DNA manuscript to the published copies can be glimpsed. One graphic insight is offered in the electron micrograph of *3*. Grouped along a strand of DNA are two feathery structures. These are formed of many small strands of the related molecule, RNA. RNA transcribes one meaningful segment of the nuclear DNA in order to carry its information—which is, at least in simple cases, the recipe for a protein—away to the molecular mechanisms outside the nucleus where protein synthesis occurs. Many RNA strands are being made here all at once; only the longest ones are complete. The example here is quite a specialized one, although transcription as a process is universal in life.

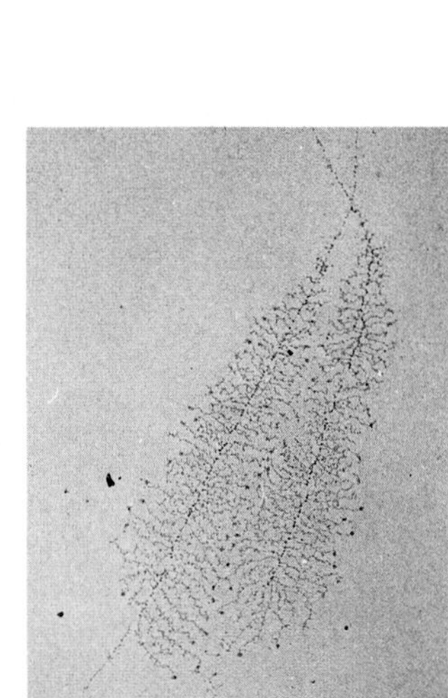

3 1 μ

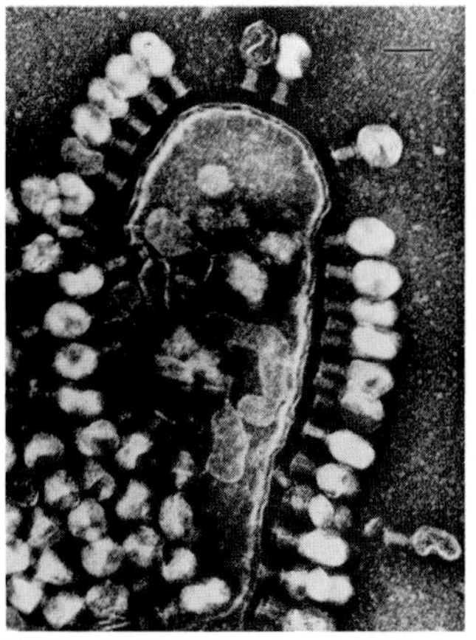

2 1 μ

We do not know just how the DNA tape is packed within the chromosome of the animal cell. One step of the cunning scheme is indicated in the electron micrograph in *4*. A DNA strand from a chromosome has been teased out into a long string. Along the string are little beads of a specific protein. It is believed that the DNA can wind around the protein beads like so many spools. The spools are, in turn, stacked close to form a supercoil, which can somehow coil and coil again into the intricacies of the chromosome: In the end, a few centimeters of DNA are readably stored on such a micron-sized scroll of protein.

The genetic molecule of *1* does not belong to any cell. It is the description of a virus, a kind of genetic highjacker, not itself alive. This DNA cannot replicate unless it is able to seize command of the machinery of the right living cell. There it can take over, to dictate its own replication and the manufacture of specific proteins that arm and protect it. The combat photograph of *2*, another electron micrograph, shows a number of virus particles attached to their host, a bacterial cell. Their protein coats, each a kind of tape cassette, are shaped to fit structures in the cell wall. Upon contact, the springy protein can inject the virus's DNA tape into the cell, where it redirects the bacterial machinery, toward synthesis not of cell material but of more virus. Some new virus heads have been made already; the cell has burst open.

Many kinds of viruses are known, each able to infect specific cells, whether bacteria, oak, or human. It is probable that viruses are not early forms of protolife, but rather genetic renegades originating during the evolution of their host cells.

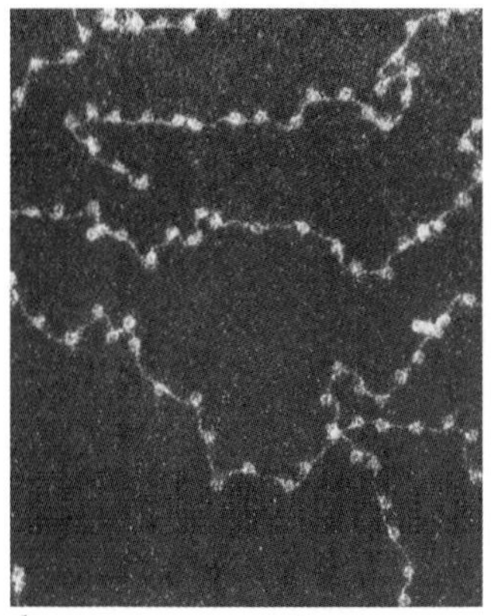

4 1000 Å

Held safely inside the cell nucleus are enormously long molecules, the coiled coils of DNA cunningly spooled and folded within this tiny space. These vital instructions are carefully duplicated at every cell division. One such thread of DNA, a few centimeters long, is stored in each of the forty-six chromosomes within the nucleus of every human cell.

THE DOUBLE HELIX

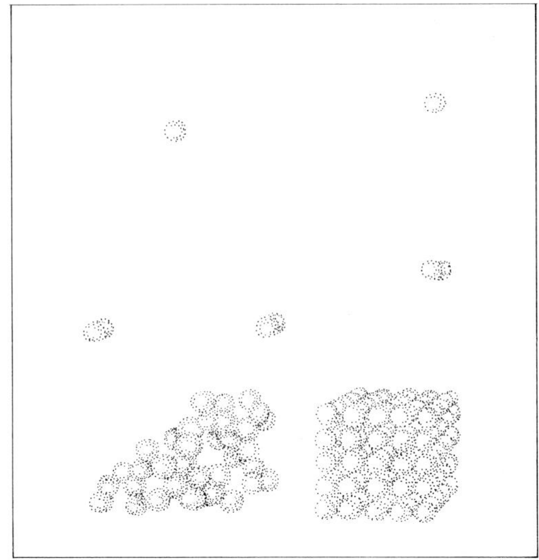

1 20 Å

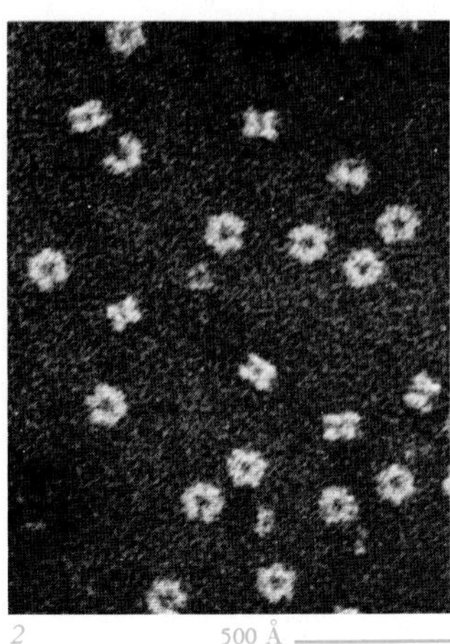

2 500 Å

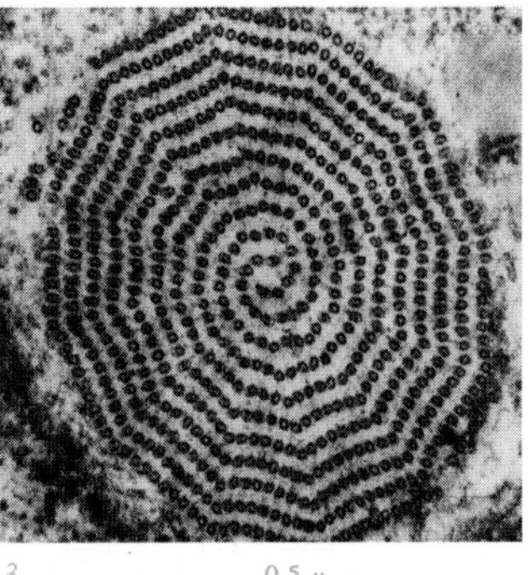

3 0.5 μ

Within the last few years direct if crude images, as in *2*, have bridged the gap between our assisted vision and the molecular models like that of DNA, which physics and chemistry had steadily elaborated from a complicated chain of inferences. The quality of such inferred molecular models can be judged by a look at a well-founded and long-familiar example, even though it has not yet gained direct visual support. This is the molecular model of a typical gas, a liquid, and a solid, drawn in *1*. The simple molecules that make up an ordinary gas like room air are well spaced, by five or ten diameters; they vibrate and rotate as they freely fly, moving at random with the speed of sound until they strike, to rebound again and again from each other and their confining walls. (And the walls themselves are always in tremor.) Close-packed, interacting, and well ordered, the same molecules can bind into a crystalline solid. There the unceasing motions are restricted, vibratory. In the intermediate liquid form, the molecules jostle past each other, only transiently bound. Such a model makes understandable the easy compression and expansion of a gas—the molecules on average simply find more or less space for their flight—and the strong resistance of a solid or liquid, where compression requires actual atomic deformations.

Notice the profligate number of moving molecules that make up even a small sample of matter. A cube one hundred angstroms on an edge holds about 10^5 simple molecules in the solid phase, and a couple of hundred in a gas like normal air. It is these large numbers that allow the variety of material properties, even when the architecture is as repetitive as that of a generally orderly crystal.

Self-duplicating DNA is about as elaborate a piece of molecular architecture as we know well, more subtle than any tape recording, whose form and function it somewhat recalls. Even the shorter DNA threads of bacteria are built of atoms by the hundred million. But much smaller molecular structures that perform simpler vital functions abound in living cells. The image *2* shows a few poorly seen identical molecules of one specific bacterial protein; each of these forms is a complicated coil wound of some fifty thousand atoms.

The molecular biologist is coming to picture living cells as systems in which a kind of elegant engineering assembles the cool symmetries of certain molecules into complex functional mechanisms. The photo in *3* shows an example far from genetic or chemical function. The pattern is the cross section of a bundle of molecular tubes that make one microspine of the scores radiating from the central body of an aquatic microorganism. The structure is tens of microns long; it is used for feeding and perhaps for motion.

I had decided to build two-chain models. Francis would have to agree. Even though he was a physicist, he knew that important biological objects come in pairs.

J. D. Watson,
The Double Helix, 1968

In this close-up the DNA is seen as a long twisted molecular ladder, the double helix. The individuality of the organism is held in the running sequence of the differing rungs. Tha chemical message is spelled out at great length in a molecular alphabet of four letters. One alphabet serves all life, but the tale retold in every cell of the body differs from individua to individual. The two rails of the ladder come apart during cell duplication, each to ac as a template for one complete new copy of the ladder of rungs.

MOLECULAR MODELS

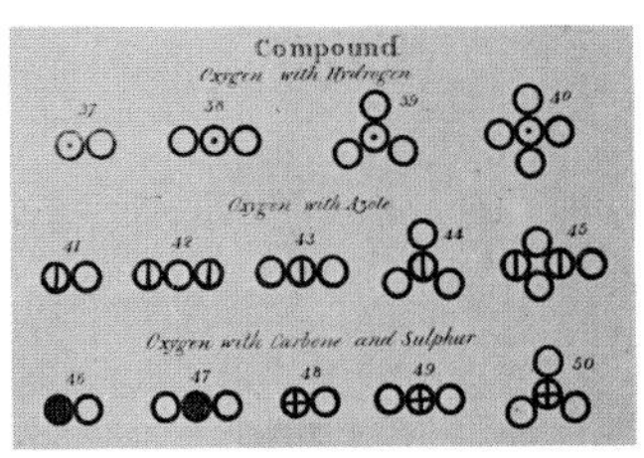

1

John Dalton displayed these pioneering explicit models *1* for some simple molecules before 1810.

Chemical logic ripened beyond the speculative atomic ideas of the Ionian philosophers in the light of chemical experience. The notion of the elements became as concrete as the realization that just about anything we eat can char to the same single black stuff. The gradual entry of relative weights and volumes as secure, simple laboratory data led to the familiar notion of atomic partnerships, summed up in commonplace formulas: H_2O for water.

By the 1870s, though the chemists still had little idea of the size of their molecules, they had new firm arguments. They knew the atomic partners linked into molecules and they knew many spatial forms, based on the logic of the possible compounds that result from modifications of one starting substance. Agreement with geometrical symmetry became clear, and the chemists boldly pictured their molecules (in the mind's eye) as well-formed bodies moving in space before physicists or philosophers would much assist or accept their work.

Today the model of a molecule is more than a teaching device, even more than an approximate description of the tiny reality. It is itself a powerful research tool; the double-helix discoverers made much use of molecular models of cardboard, wire, and tin. The models had to fit all the data under clear rules. The rules drawn for a wide group of stable compounds—they never exhaust all the possibilities—reflect established facts: Atoms seek to bond with fairly constant distances between each atom and specific neighbors; definite angles must be held between certain bonds; atoms must pack without interfering contact, each demanding specific room for itself.

All these models are partial, abstracted. They do not model the weights, nor do they mimic the electronic substance of the atom, nor can they share its curious surface properties. They do not move in that incessant interactive dance of the small; they do not vibrate or rotate, even when their joints are free to do so. But a good model does map some permitted instantaneous form of the invisible molecular structure.

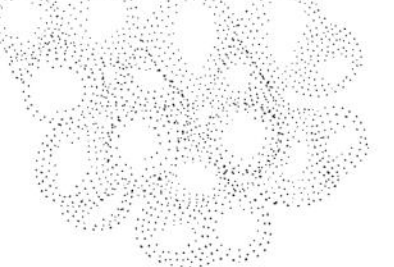

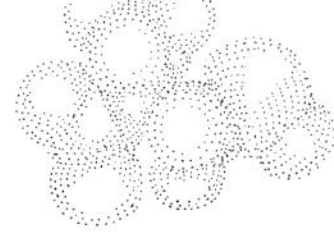

2 10 Å

Modern working models of some familiar substances are drawn in *2*. Water enables life; the two larger molecules are glucose and cysteine, building blocks—monomers—of long organic chains, no small atomic partnerships but corporate bodies, the polymers within life.

Glucose is an abundant simple sugar; cysteine is one of the two dozen amino acids that among them build all protein coils.

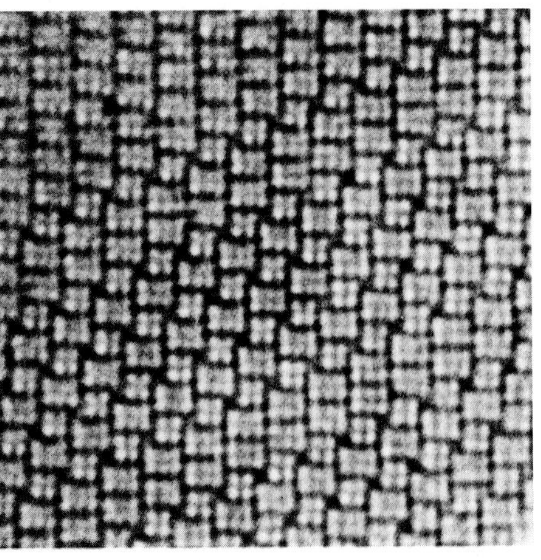

3 100 Å

The atomic array of one crystalline solid is directly imaged in *3* under about as great a magnification as we now can handle. The limitations of the technique require the use of rather refractory substances; this is a very thin crystal of niobium oxide, seen with some of its defects through the transmission electron microscope. The tapestry of atoms is complex but clear.

10 angstroms **1 nanometer**

These building blocks are molecular typography, the letters of the genetic message. It is their particular order that spells out the long text. The forms are chemical patterns, the ordinary stable structures of bound atoms, themselves indifferent to life. The central carbon atom is bonded to three visible hydrogen atoms (and to another atom that lies behind). A similar linkage might well be found abundantly among carbon and hydrogen atoms drifting in the cold thin clouds of interstellar space.

THE ATOMIC SURFACE

"Not exist? Not exist! Why, I can see the little buggers as plain as I can see that spoon in front of me!"

Ernest Rutherford to Arthur Eddington, when once the latter was musing aloud at table as to whether we would ever come to know electrons for more than mental concepts.

The forces that bind atoms in specific ways to form the motley coat that is ordinary matter act on the fuzzy surface of the atoms. All atoms share a common modular structure and a single structural principle: They are clouds of identical electrons grouped around tiny attractive centers. The surface of every atom is a pattern of electrons in motion; special symmetries are present, and electrical forces govern the motion. It is a quantum motion these binding electrons execute, under laws the consequence of this minute scale; it is not to be visualized as a simple orbit, like the track of a planet around the sun. Rather, it is inherently probabilistic. Though successive electron positions remain untrackable so long as the atom is stably bound, the resultant charge pattern is steady and sure. Shared electrons contribute to regions of increased negative charge, and the positively charged atomic cores attract those negative regions to bind the molecule.

Moreover, it is the light emitted, absorbed, or scattered by these very same exterior electrons of the atoms that determines everything visible of matter, from the glow of flame to the blackness of ink.

Since all atoms are assemblies of electrons, they share strong family resemblances. The hundred-odd different chemical elements—each element is one atomic species—are thus not really distinct. That insight is a century old: It is codified in the body of lore we call the periodic table.

19.0 Fluorine yellow intensely corrosive gas forms a strong acid with hydrogen	20.2 Neon inert gas forms no compounds	23.0 Sodium soft silvery metal reacts explosively with water to form an alkaline solution
35.5 Chlorine greenish corrosive gas forms a strong acid with hydrogen	39.9 Argon inert gas forms no compounds	39.1 Potassium soft silvery metal decomposes water to form an alkaline solution
79.9 Bromine red-brown reactive liquid forms a strong acid with hydrogen	83.8 Krypton inert gas one or two unstable compounds known	85.5 Rubidium soft silvery metal decomposes water to form a strong alkaline solution

1

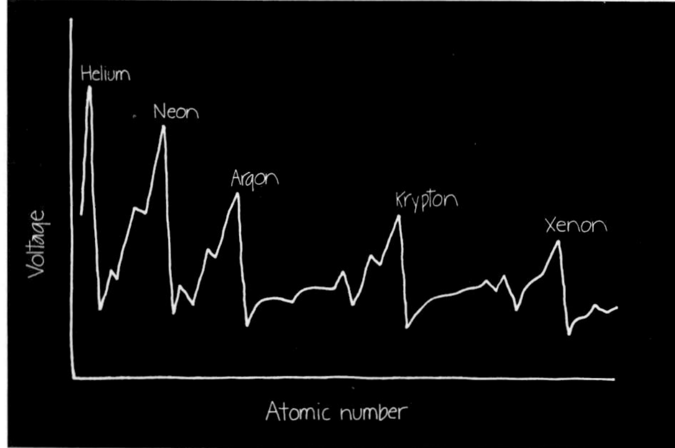

2

We tabulate in *1* some of the qualitative chemical properties of a few selected elements. They are loosely ordered by relative atomic weight. The chemical families are plain enough; resemblances appear again and again, varying not smoothly but periodically with increasing weights of the atoms.

The same periodic effects are expressed differently in the graph of *2*, not in the qualitative terms of the nineteenth-century chemist amidst beakers and acids, but in the quantitative style of the atomic physicist of the 1930s, the lab all vacuum tubes and electrical measurements. What is plotted is the minimum voltage required before an electron beam becomes just energetic enough to knock out additional electrons from the atoms of the target gas as it streams past. The periodic maxima and minima are plain; the diverse unity of the elements is again displayed.

10^{-10} meters

1 angstrom

The quantum laws of atomic scale require a description of electron motion that is more subtle and less sequential than for the moving particles of ordinary experience. Accordingly, the dot texture shown does not map individual electrons; instead, it suggests the cloud of electrical charge the electrons paint out during their symmetrical but untrackable quantum pattern of motion. In that cloud the surface electrons are shared by the bonded atoms.

The atom's innermost electrons usually dwell in splendid isolation. They move under so strong an electrical influence from the nuclear core close within that the ordinary world hardly affects them. They execute their spherical dance almost unmoved by flame or shock or chemical attack. Only extreme circumstances (perhaps in lightning or under radioactive conditions or within the sun) will much modify their symmetrical motion. They do not interact with visible light; such light cannot supply sufficient energy to knock one loose.

For most atoms beyond the very few lightest ones, the innermost electron shells can emit and absorb only those radiations we know as x rays, each interaction 10^3 or 10^4 times more energetic than that of visible light.

The striking photographic shadow cast by x rays, reproduced in *1*, is one of the first medical radiographs made in America, in February 1896. The lead pellets (dense with electrons) embedded in the man's hand by a shotgun accident are clear. Wilhelm Röntgen discovered x-rays by chance late in 1895; before 1914 they had become the probe of the innermost atomic electron shells. Fast-moving electron streams within the high-voltage x-ray tube can disturb the deep-lying electrons of the target atoms; x rays are emitted, and these can be absorbed in turn within the inner shells of other atoms.

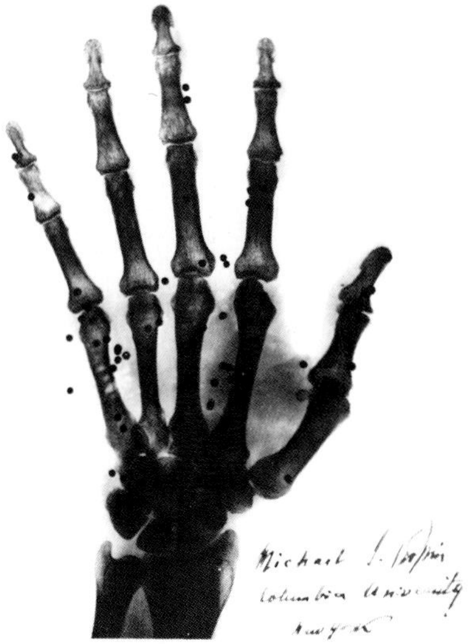

1

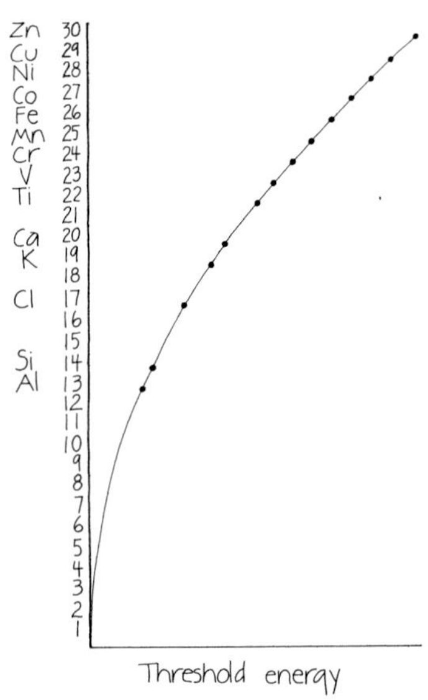

2

In the graph *2* is summarized a study of the highest x-ray energy an element can absorb strongly. This energy measures the response of the innermost atomic shell in each atom to incoming x rays. The heavier the atom, the higher-energy x rays it can absorb. The smooth relationship is persuasive. The rank order of the element by x-ray energy absorbed musters all the atomic species in an even more regular way than does the relative atomic weight. That unique label is called the atomic number; it amounts to a count of the total number of positive charges that bind the innermost electrons. That is the same as the total number of electrons held in the complete electrically neutral atom. The atomic number supplies the most rational label for any element; periodic relationships and the interactions with light and with radiation in general are all better ordered under such a scheme.

0.1 angstrom **10 picometers**

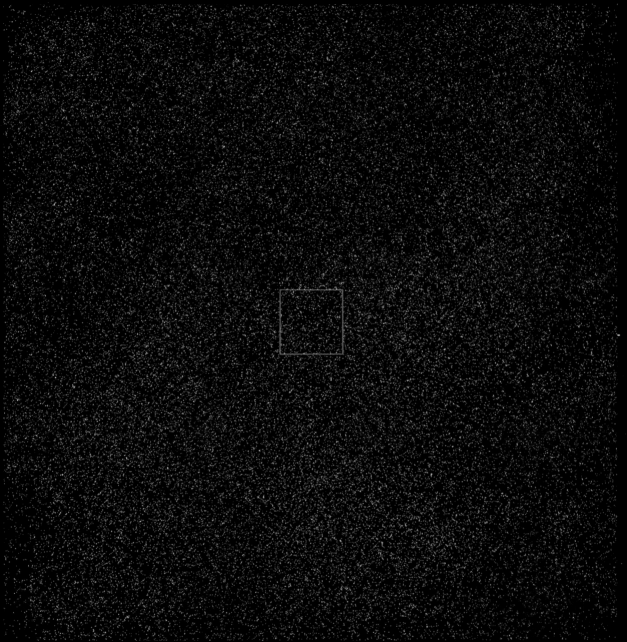

Now we are among the two innermost electrons of the carbon atom. They mark out in their dance a neat sphere of electric charge. The four outer electrons of carbon can come and go, whether in flame, in diamond, or in DNA. But these inner electrons remain indifferent to ordinary experiences, which cannot disturb their seclusion; they respond only to the nucleus within.

It is extraordinary how small is the atomic nucleus compared to even the atom. For not only does the nucleus hold all the positive charge of the atom, it is also the site of the weight of every atom. The electron cloud plainly occupies the bulk of an atom, but the electrons in an atom together contribute relatively less to its weight than the postage stamp adds to a book sent parcel post. The nucleus is massive and tiny; within it intense binding forces are at play. As the electrons, all identical, supply all the negative electric charge in ordinary matter, so the protons comprise all the positive charge. Electrical unlikes attract; the proton clusters that form the central nucleus thus bind the whole atom. Each element is one species of proton cluster, its defining integer the head count of its protons, called its atomic number.

Ordinary hydrogen is the lightest of all atoms; it has weight very near 1 on the scale in which the carbon-atom standard is 12. Hydrogen has unit atomic number as well: one proton. The hydrogen atom consists of a single positive proton binding a single negative electron. Carbon has weight 12, yet its atomic number, its proton count, is only 6. That is usual enough: For example, the heaviest naturally occurring atomic species, uranium, has a weight of about 238, but an atomic number of only 92. Something adds weight to the nucleus, enough more or less to double the weight in protons, but contributes no charge at all. In 1932 that heavy neutral particle was found; it is called the neutron.

A free neutron is really a strange sort of chemical element, atomic number zero, weight one. Without charge, it can hold no atomic electrons at all: It occupies only its own tiny nuclear volume, with none of that fluffy electron-cloud bulk; it is without ordinary chemistry. We all know of it; it is the reagent of nuclear transformations *par excellence*, for it combines on nuclear contact with almost every other nucleus we know, to release a large amount of energy—atom for atom a millionfold the energy of chemical reactions like ordinary combustion. The end result is alchemy: the transformation of one atom—that is, one element—into another, typical during deep nuclear changes. (It isn't economical to make gold from lead in such a way, but it can be done.)

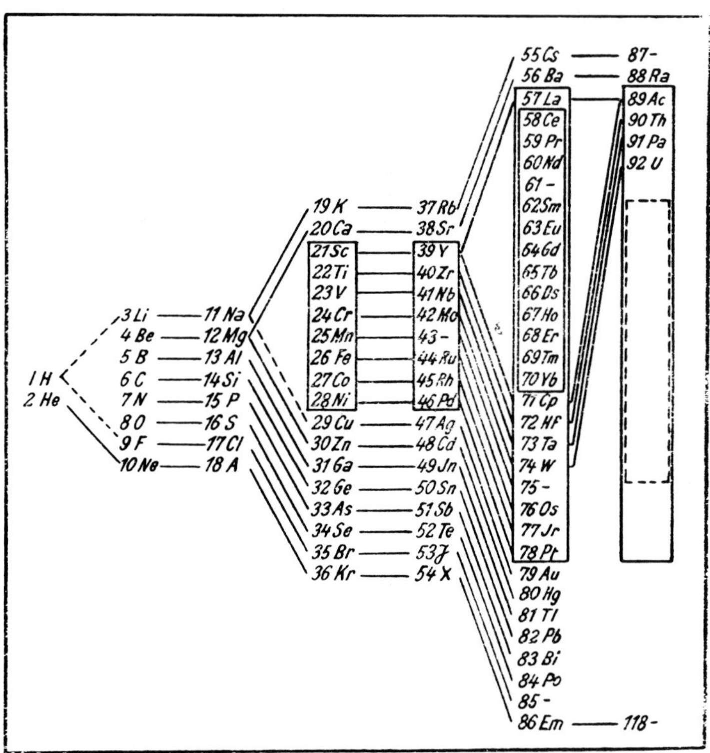

The system of the elements as presented by Niels Bohr in 1923 according to his early theory of atomic structure. The table shows the chemical symbol for each element and its atomic number. Lines connect the elements of the same chemical family. By now a dozen elements with atomic numbers beyond 92, uranium, have been synthesized through successive nuclear reactions; all are unstable.

10^{-12} meters

1 picometer

The compact core of the atom begins to appear. The balance of atomic force is set by this nucleus, whose strong electrical attraction binds the electron dance. To bind six negatively charged electrons, exactly six positive protons must cluster within the nucleus: That number (the atomic number) defines the element carbon. We know about a hundred distinct species of these tiny proton clusters, the elements: Modular but diverse, they determine the material universe.

An element is one chemical species, a class of nuclei with a particular number of protons. Carbon atoms have six protons, hydrogen atoms only one, and so on. That proton count defines the chemistry of the atom. The neutrons add weight and play a key role in the central nuclear structure, but they hardly affect the properties of the external electron cloud.

It is the proton count that characterizes the chemical element, be it carbon, chlorine, or gold. Whether there are more or fewer neutrons linked with the protons in the nucleus does not much affect its chemistry. Two atoms with differing neutron numbers but with the same proton number are called isotopes: The root of the word means "in the same place" within the periodic table. Most natural samples of elements are isotope mixtures; every nucleus has one and the same atomic number, but several different neutron numbers are present.

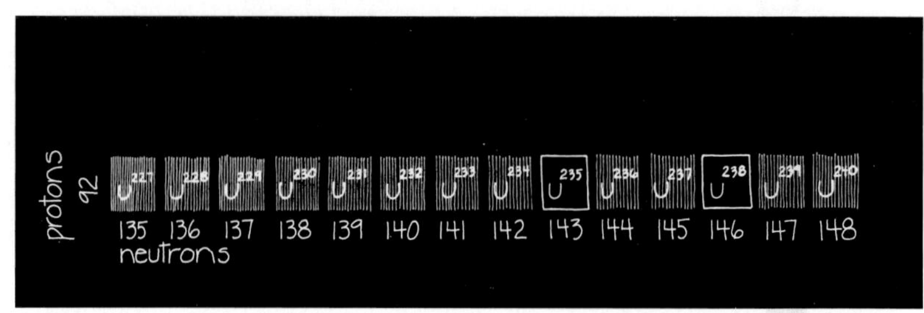

2

The small excerpt *2* of the heavier end of the isotope chart selects out the isotopes of uranium, the heaviest natural element. Notice how the neutron–proton balance has shifted. The rare natural isotope U^{235} is industrially separated from uranium by very expensive processes because it can support a neutron chain reaction, whereas the much more abundant and chemically very similar natural isotope U^{238} cannot. No isotope of uranium is really stable, but a few of them do endure, slowly decaying over the ages, for the lifetime of the earth.

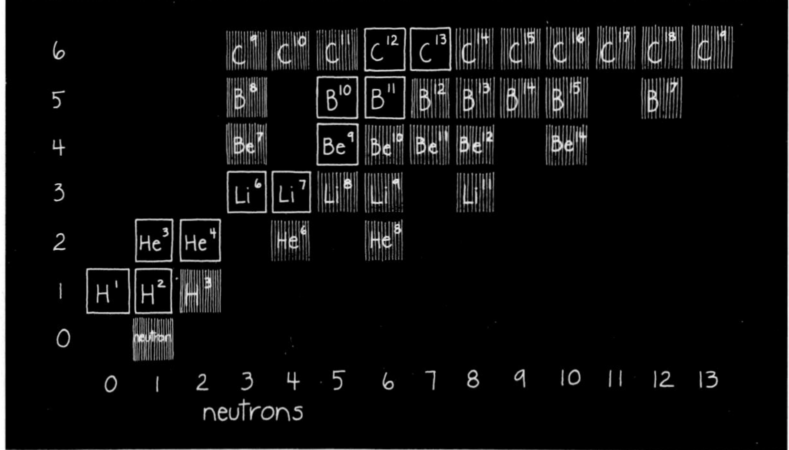

1

In the chart *1* are arranged the known isotopes of the first few elements, through beryllium, which has four protons. Most of these nuclei are synthetic, transient, unstable, radioactively decaying in a time short compared to the age of the earth. The more stable natural isotopes have somewhere near equal numbers of neutrons and protons; this balance shifts slowly to a neutron surplus as the nuclear weight grows. By now we know some two or three hundred stable isotopes among all the known elements, and ten times as many unstable ones. Their study is the subject matter of nuclear physics and chemistry, a kind of inner energy-rich analogue to the electronic physics and chemistry of the outer atom.

0.1 picometer **100 fermis**

We see clearly the minute and massive kernel of this particular carbon atom. Its close-packed nuclear components are in vigorous quantum motion, but here the motion is profoundly restricted and fluidlike. Bound by nonelectrical nuclear forces of terrible strength but of very limited reach, the six neutrons and six protons seem to touch. With twelve nuclear particles, this nucleus is dubbed carbon-12: The most common isotope of carbon, it is the modern standard of atomic weight.

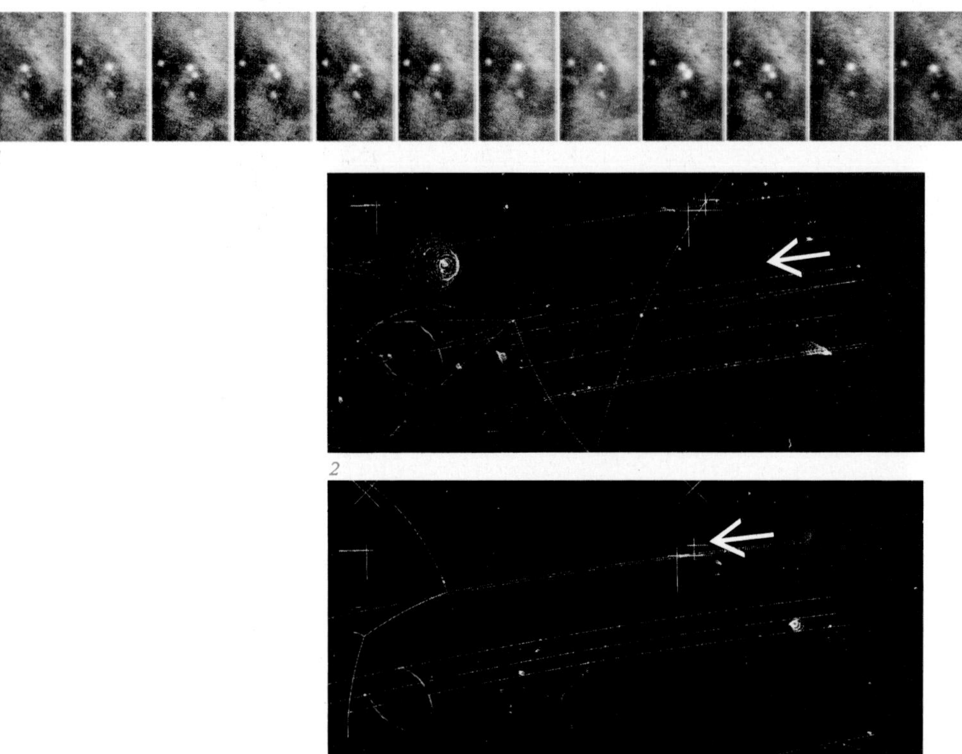

1

2

3

These frames from a moving picture show not a nuclear particle but a distant star. It is a minute star a thousand times brighter than the sun, flashing out thirty times a second as it spins, its beamed light crossing and recrossing the field of view like the beam of a rotating light in a lighthouse. In fact, it is a star made of one huge nucleus. Here not a few hundred nuclear particles but about 10^{56} of them have joined, under mutual gravitational forces, to constitute a chunk of nuclear matter as massive as the whole sun yet no larger across than a small city. This is not the only context in which we begin to find unexpected links between the worlds at very different powers of ten.

The flash-lit photos look through a thick window into the depths of a tank of liquid hydrogen, momentarily on the verge of boiling. The white lines that seem etched on a black ground mark the tracks of high-energy charged particles. The particles streak through the fluid, tens of centimeters or more, and a trail of tiny visible bubbles then begins to grow around the disturbed atoms left along the track. Uncharged particles disturb so few atoms as they pass that they leave no visible wake.

Note the pairs of particles that appear together at one spot out of the blackness *2*. These are electron and positron pairs, matter made anew out of the energy of unseen radiation. If one follows the Vs backward, they point to the place where the incoming charged particle first interacted with a proton of the fluid to produce charged particles and the unseen neutral radiation that made the two pairs. The two members of each pair curl oppositely, the orbits imposed on positive and negative electric charges by the strong magnetic field in the chamber. (The incoming particles are unstable positive pi-mesons.)

In *3* an incoming meson strikes a stationary proton of the fluid, to project it forward and down. That proton proceeds to carom off another resting proton, which itself moves on a little distance. The right angle seen at the short prong is the sign that the two collision partners have about the same weight. Meanwhile, the incoming meson moves off upwards.

Careful piecing together of this sort of subnuclear billiards in many and varied examples establishes the properties of the particles and their interactions.

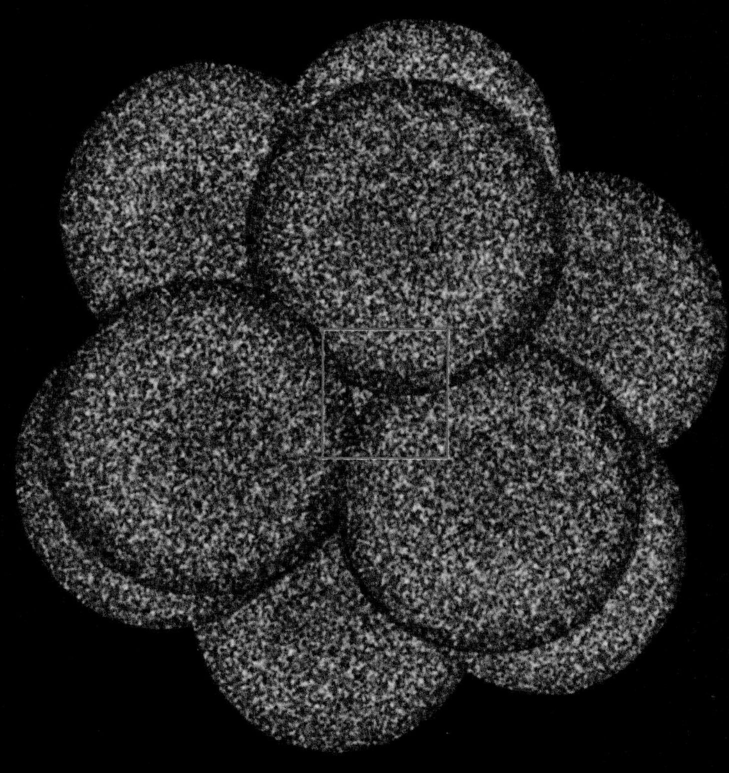

A transient view of the eternally dancing structure of stable carbon-12. Those neutrons and protons that join to form it are universal nuclear modules. Protons are found free as natural hydrogen; neutrons can be set free by energetic nuclear reaction as in the fission of uranium. Study of these particles as independent objects has revealed one more analogue to chemistry: They too react upon collisions at high enough energy to produce a host of new particles, mostly transient ones.

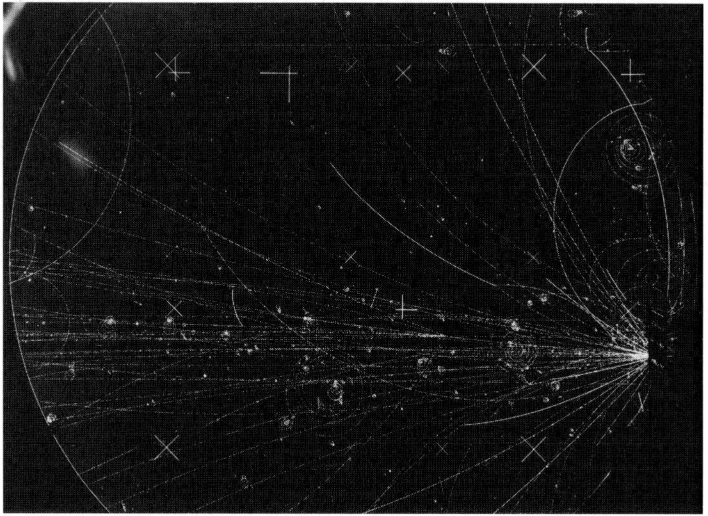

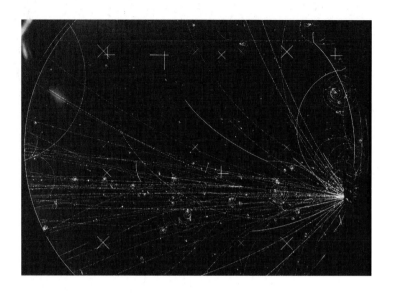

This track photograph records a complex high-energy multi-particle event. Here we transcend the stable world of protons, electrons, and radiation; we see a whole fan of new and transient particles (the lighter trails). Most of these can be thought of—though firm proof is not yet to hand—as strongly bound combinations of particles never yet seen on their own: the quarks. Quarks bear several novel kinds of charge, analogues to electrical charge though quite distinct from it. Three specific quarks combine to build a proton, three different ones a neutron; quark pairs constitute the well-observed unstable particles called mesons. Other quark combinations make more kinds of transient particles, such as the tracks shown here; plenty of protons leaving heavy trails are seen as well.

All the tracks here arose from one initial event: A single heavy nucleus within the gold foil (to the right) was struck and fragmented by one incoming particle.

The particle physicists enter the eighties proud of the order new-won during the last decade. The modular world of all the matter and radiations we know currently appears to be built up of three small families of particles, not yet fully verified. First are the strongly interacting nuclear particles, proton and neutron and a whole transient subnuclear world, all made of a few quarks in combinations. Second are the stable electron and neutrino, and a few kindred particles, all with simple structure and comparatively weak interactions. Finally, there are a number of new analogues to the photon, which itself mediates the electrical and magnetic forces by quantum exchange among charged particles. Each of the known fields of force has its specific set of mediating particles.

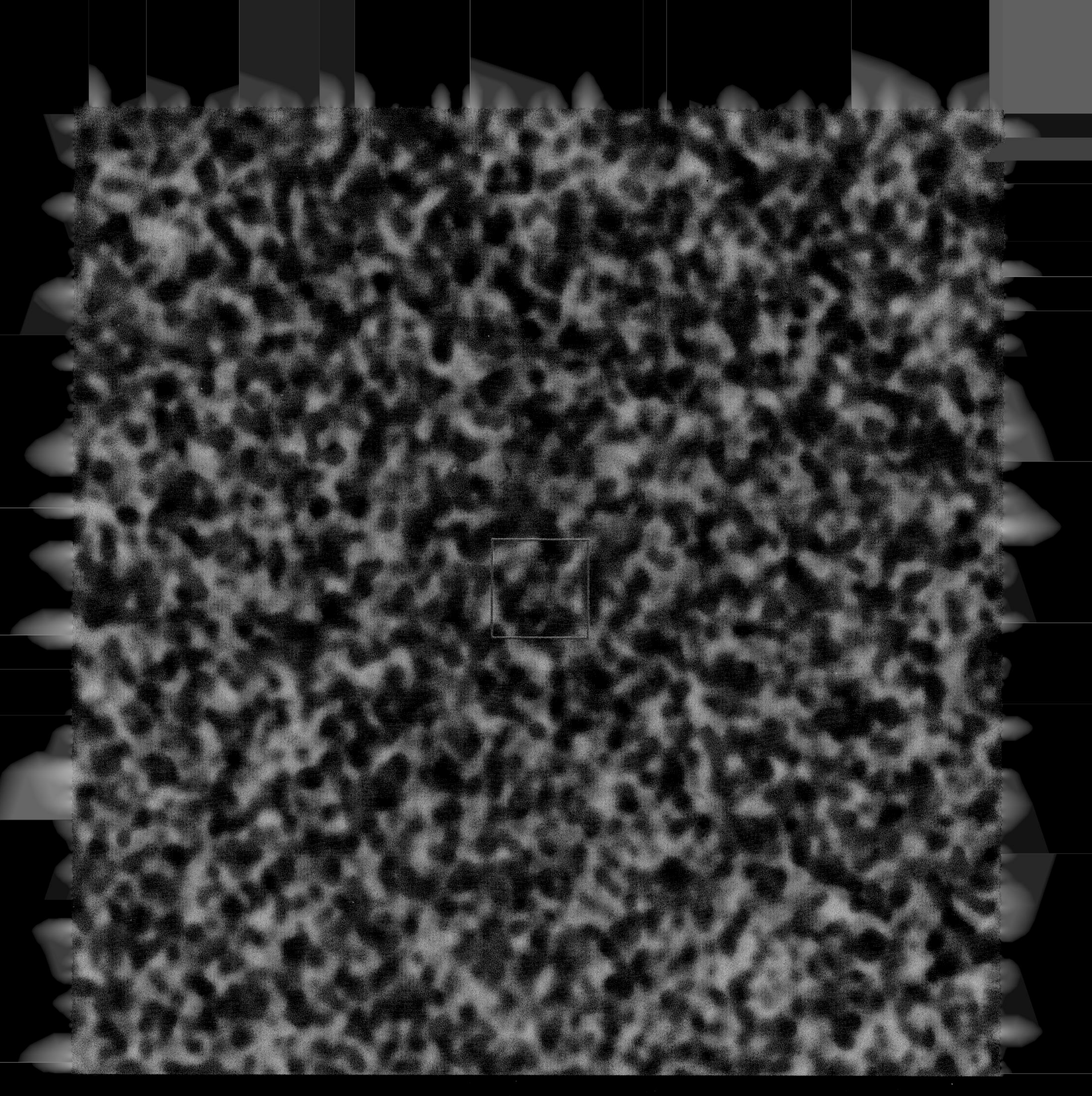

Even the proton has its inner structure, symmetrical, shifting, again untrackable. Here stil stronger forces operate at still shorter ranges. These arise among fast-moving quarks ir intense interaction. The pattern of colored dots is no photo but an abstract symbol of the

Some say novelty may lie just next door, at 10^{-17} meters. But there is a serious view nowadays among the theorists that quarks and their associated particles more or less represent everything to come—until we zoom down, if ever we do, to a distant power, 10^{-31} or 10^{-32} meters. There lies predictable new structure, they say, a new and yet more remarkable particle world.

Time will tell.

He shewed me a little thing, the quantity of an hazel-nut, in the palm of my hand; and it was as round as a ball. I looked thereupon with eye of my understanding, and thought: What may this be? And it was answered generally thus: It is all that is made.

The Anchoress Julian
of Norwich, about 1400

10^{-16} meters

0.1 fermi

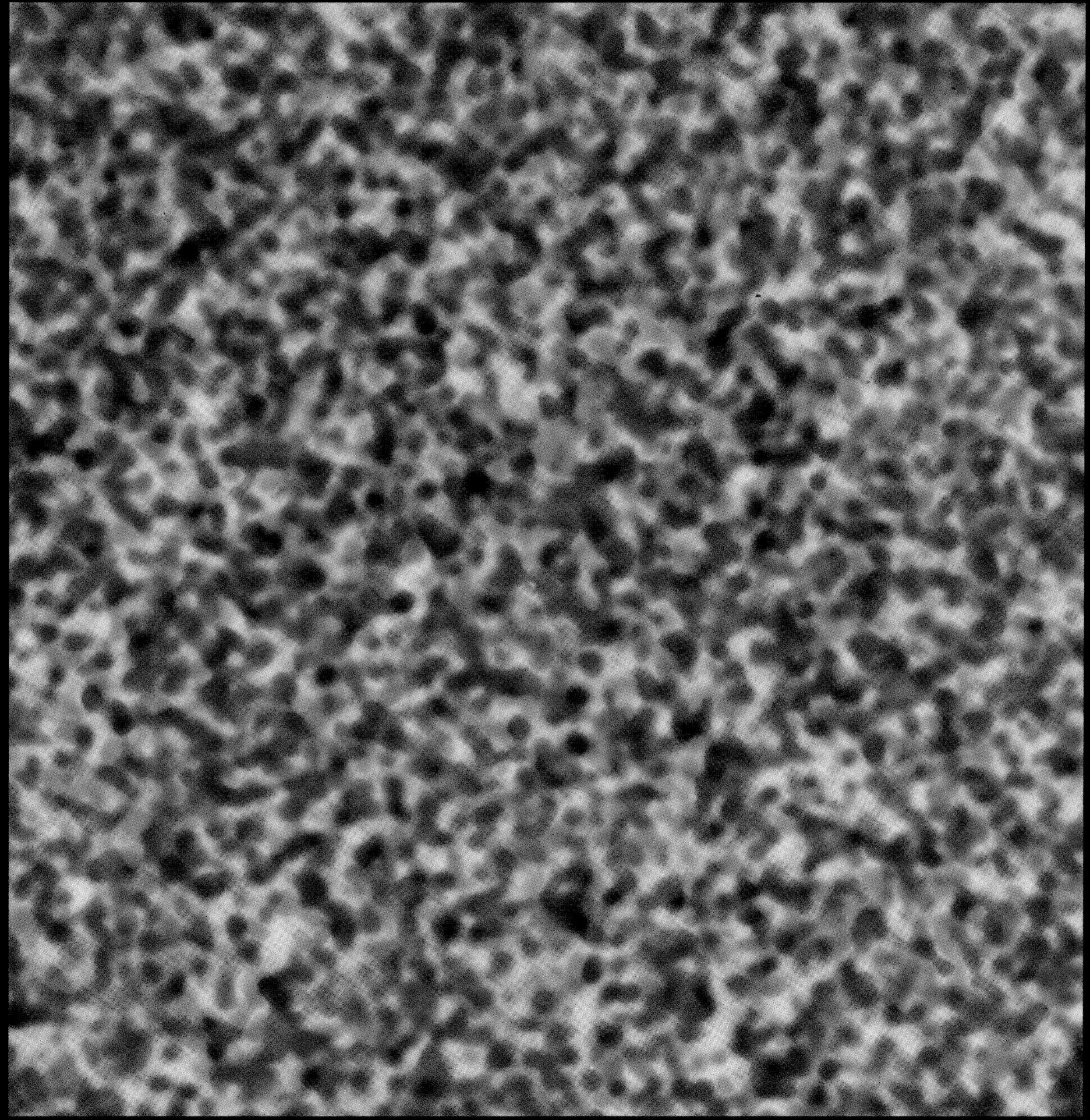

What will we see, and what will we come to understand, once we enter the next levels?

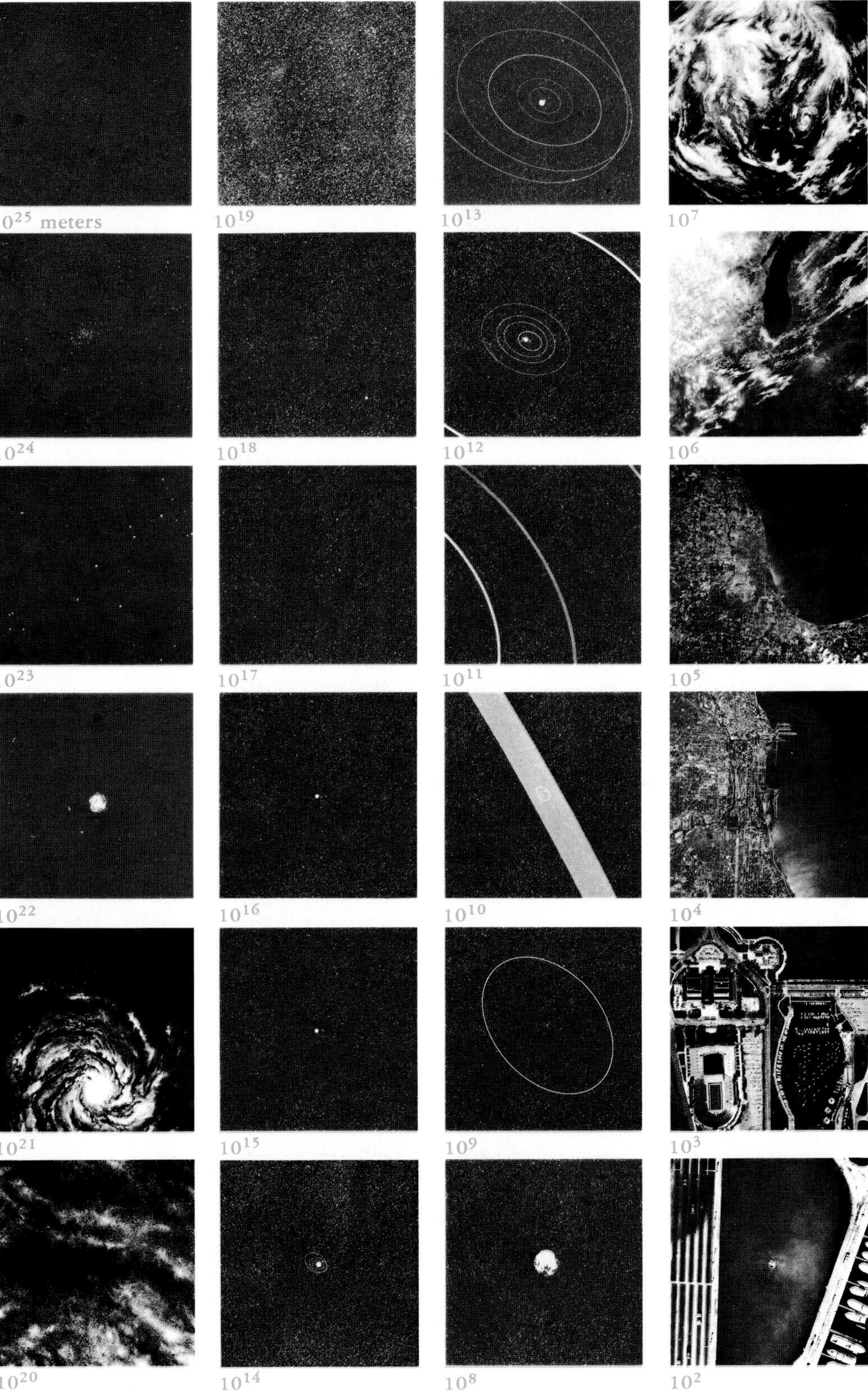

10^{25} meters
10^{24}
10^{23}
10^{22}
10^{21}
10^{20}
10^{19}
10^{18}
10^{17}
10^{16}
10^{15}
10^{14}
10^{13}
10^{12}
10^{11}
10^{10}
10^{9}
10^{8}
10^{7}
10^{6}
10^{5}
10^{4}
10^{3}
10^{2}

10^{1}

10^{0}

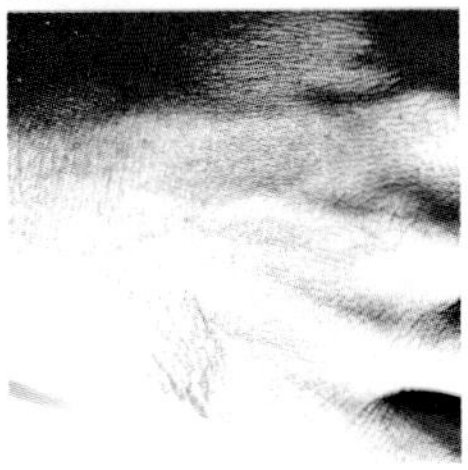

10^{-1}

10^{-2}

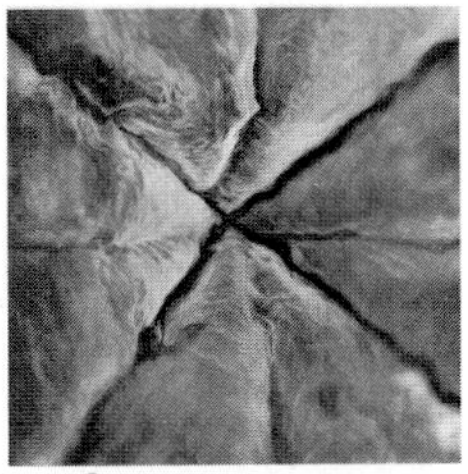

10^{-3}

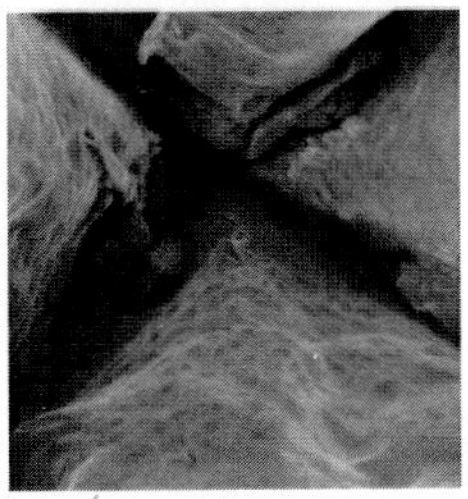

10^{-4}

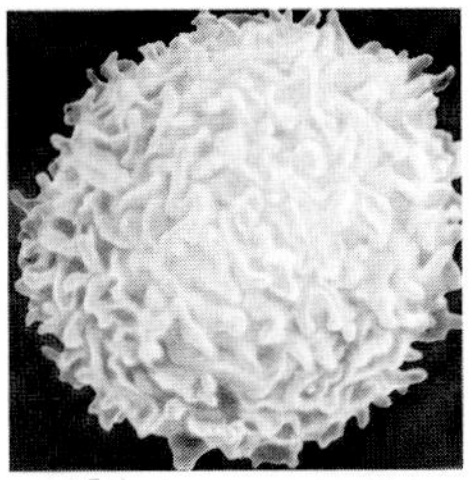

10^{-5}

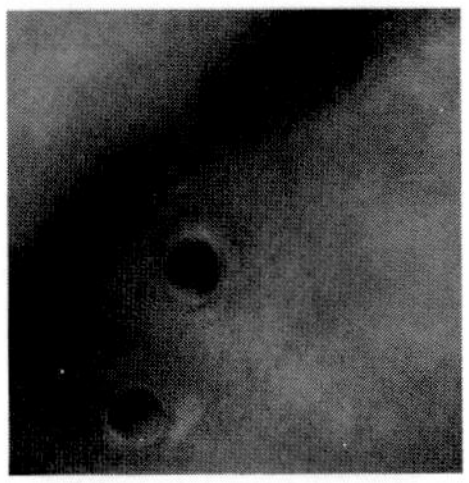

10^{-6}

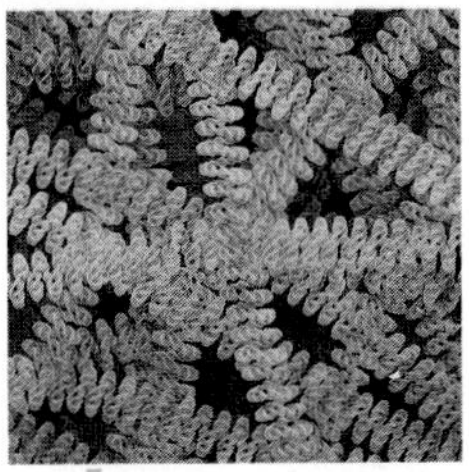

10^{-7}

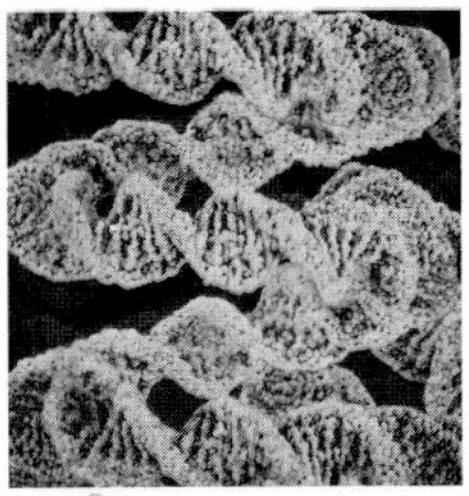

10^{-8}

10^{-9}

10^{-10}

10^{-11}

10^{-12}

10^{-13}

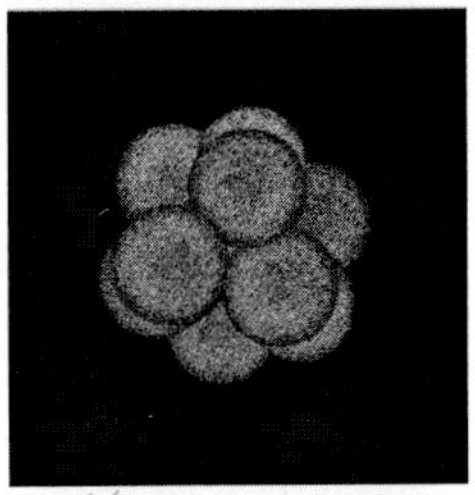

10^{-14}

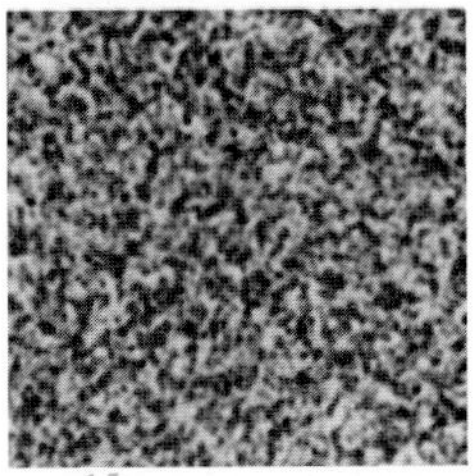

10^{-15}

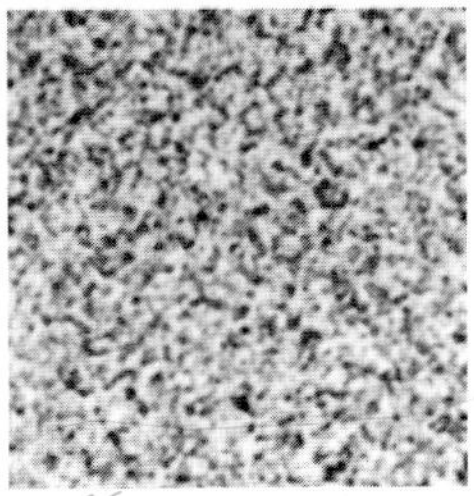

10^{-16}

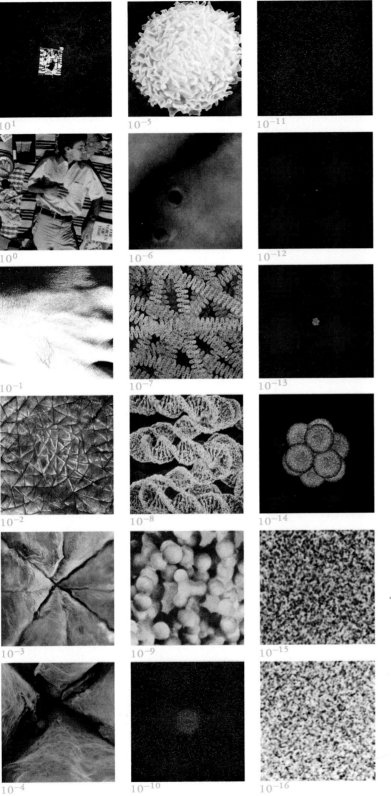

IN COMMENT

RULES FOR A JOURNEY OF MIND AND EYE

These rules for the reader express the logic that guided the construction of the entire sequence of forty-two images.

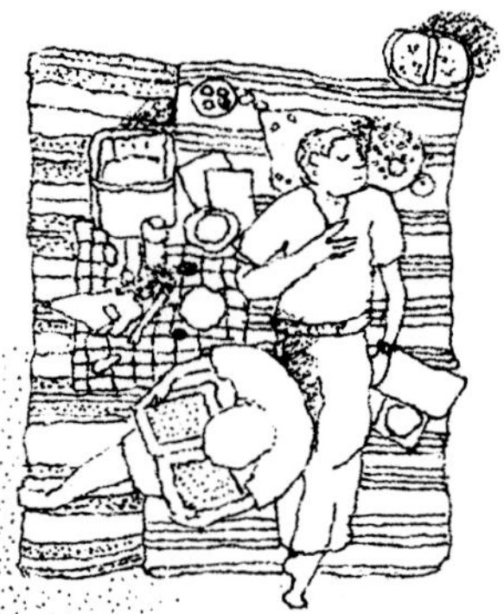

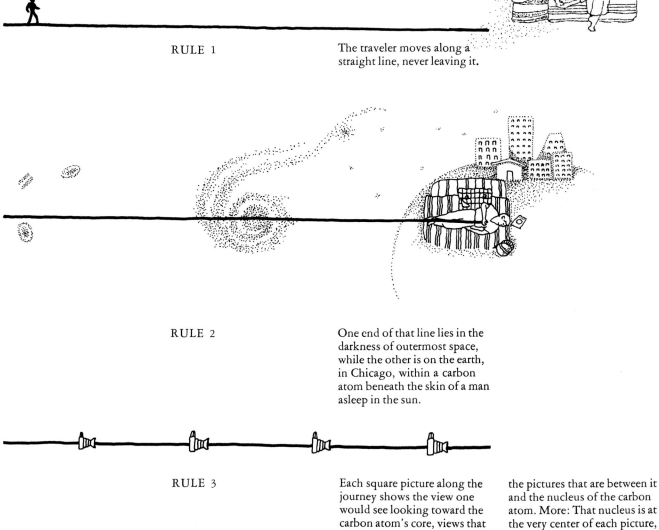

RULE 1

The traveler moves along a straight line, never leaving it.

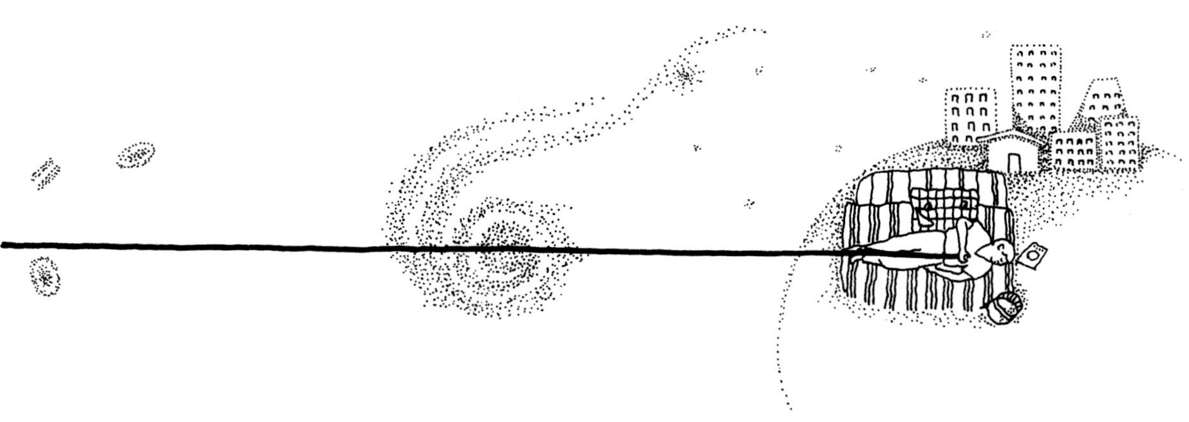

RULE 2

One end of that line lies in the darkness of outermost space, while the other is on the earth, in Chicago, within a carbon atom beneath the skin of a man asleep in the sun.

RULE 3

Each square picture along the journey shows the view one would see looking toward the carbon atom's core, views that would encompass wider and wider scenes as the traveler moved farther away. Because the journey is along a straight line, every picture contains all the pictures that are between it and the nucleus of the carbon atom. More: That nucleus is at the very center of each picture, whether it can be seen or not.

RULE 4

Although the scenes are all viewed from one direction, the traveler may move in either direction, going inward toward the carbon atom or outward toward the galaxies.

RULE 5

The rule for the distance between viewpoints needs to be carefully chosen. If the steps are equal, two problems arise. To fit in the pages of our book, the steps would have to be so inordinately long—each one covering about 50 million light-years—that the first step away from our target would far overshoot Chicago, to land the traveler immediately in the realm of the galaxies. If we tried a smaller pace, say one meter at a time, then our book would have to be very long: 10,000,000,000,000,000,000,000,000 pages would do it!

We can choose a special kind of regular step that changes in size along our journey: paces in geometric progression. In that case, each step is *multiplied* by a fixed number to produce the size of the next step: The traveler can take small, atom-sized steps near the atom, giant steps across Chicago, and planet-, star-, and galaxy-sized steps within their own realms.

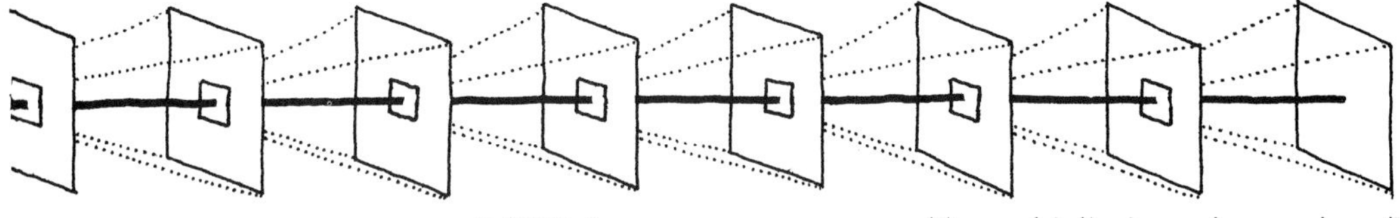

RULE 6

The multiplier in our journey is a factor of ten, bringing a tenfold change to both the step size and the field of view seen along the way.

Each image square shows a view ten times wider or narrower than its neighbor. In either direction, new information is found at each step: Outward, novelties appear as the view widens; inward, the expansion of the picture allows the resolution of more detail. That relationship is emphasized in each picture by outlining a small central square whose edges are one-tenth the width of the whole frame.

Ten is a good factor for our purposes; each new picture reveals much that is new, but the central area retained—it is only a hundredth of the whole in area—is still recognizable. And taking many steps, each of diminishing size, has a surprising consequence; it opens up the inner world of the very small.

Notice some concomitants to this scheme: An object changes its size by a factor of ten—people speak of *an order of magnitude*—with any single step; a double step brings a hundredfold change. Three steps and the scale changes by a thousand, no matter where you start or which direction you choose. When going inward, every step covers nine-tenths of the distance remaining to the heart of the atom: Like the competitors Achilles and the turtle, the traveler on our line can never quite arrive at his target.

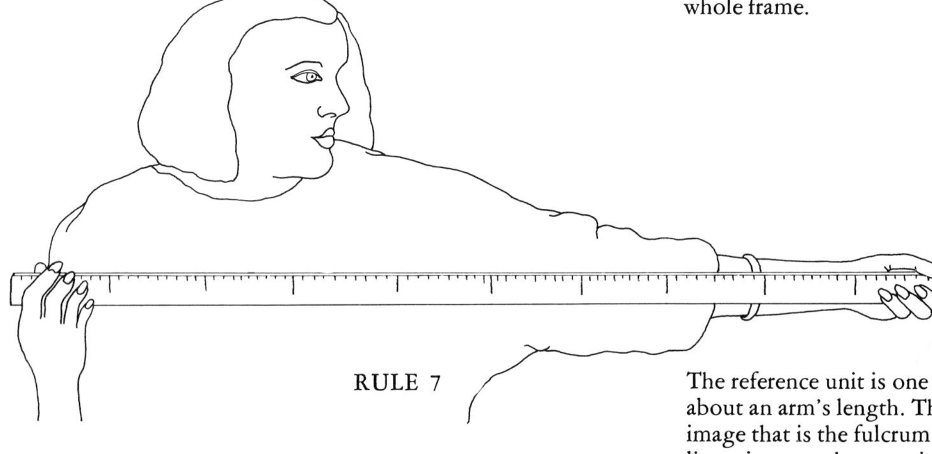

RULE 7

The reference unit is one meter, about an arm's length. The image that is the fulcrum of our linear journey shows a picnic, in a view one meter wide. The factor of ten by which the journey unfolds was chosen also because it uses the convenience of our decimal numbers and that of the metric measurement system, both based on the ten fingers of our hands.

When a number is multiplied by itself a certain number of times, the result is called that power of the number. Our journey proceeds in steps of powers of ten: The number label for each picture of our journey shows the linear size of the view in the picture square measured in meters. The number of meters is written in powers of ten notation. (A full account is given on the following page.) The picture edge and the central square provide a handy scale by which the true size of any object shown can be estimated.

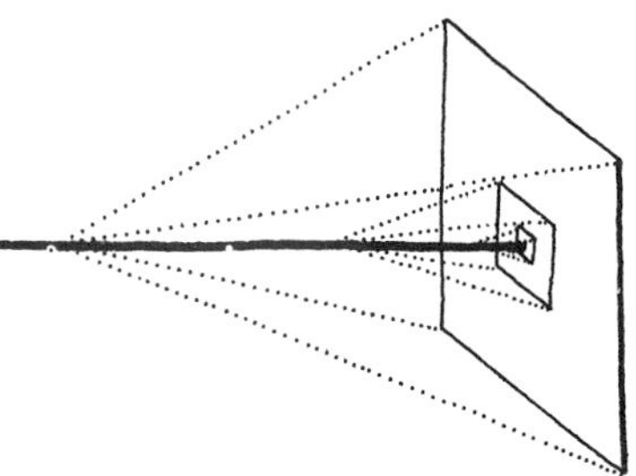

Our journey's ends are not arbitrary. For the far end, we pause at that distance where further steps would give us no new structures in space. The elusive inner target is approached as closely as our present knowledge permits. Others who come later will travel farther.

Nor is that straight route a randomly chosen line. True, the outermost portion must be typical of space; there is not much to choose from out there between the galaxies. But the inner end was placed with care to make the trip full of interest. That end lies within carbon, the most interesting of atoms to living creatures, and within DNA, the most important of organic molecules. The route takes us vertically upward, to view from above the great city on the lake shore; it is a daylight view. The time of day and year could then be chosen so that the line not only passes vertically away from earth, but also straight out perpendicular to the flat disk of the Milky Way Galaxy.

Each view looks at a volume of space that has a particular shape, a square-based pyramid; nesting within are smaller and smaller pyramid-shaped volumes. In all views while we are still limited to earth, we see no farther than the ground, the shared base of those pyramids. Once the earth becomes a floating globe, we can also see past it, to stars far, far beyond—the very stars we would see in the night sky of Chicago's earthly antipode.

Finally, this is not a journey over time. It is a set of views, all referring to one single moment in time. If we had to travel physically to take the enormous trip, we would be limited, as are all moving objects, to the speed of light. We could not make greater and greater leaps in one fixed time. Such rapid motion would distort even the measurement of time and do queer things to the light by which we see. The convention used is this: Imagine the photos were made all at one time by competent observers who agreed on when and how they would prepare their photos. (That final assembly would take a very long time.) What was done amounts to a synchronized collaboration, by preparing all the images at the right scales at one physical place.

POWERS OF 10: HOW TO WRITE NUMBERS LARGE AND SMALL

This book uses a notation based on counting how many times 10 must be multiplied by itself to reach an intended number: For example, 10 x 10 equals 10^2, or 100; and 10 x 10 x 10 equals 10^3, or 1000. Multiplying a number by itself produces a *power* of that number: 10^3 is read out loud as "ten to the third power," and is another way to say one thousand. In this case, there is no great advantage, but it is much easier and clearer to write or say 10^{14} than 100,000,000,000,000 or one hundred trillion. After 10^{14}, we even run low on names. The number written above in smaller type—the 14 in the last example—is called an *exponent*, and the powers notation is often called *exponential notation*.

It is not hard to grasp the positive powers of ten—10^4, 10^7, 10^{19}—and how they work; but the negative powers—10^{-2} or 10^{-5}—are another matter. If the exponent tells how many times the 10 is to be self-multiplied, what can an exponent of −5 (negative five) mean? The system requires a negative exponent to signal division by 10 a certain number of times: 10^{-1} equals 1 divided by 10, or 0.1 (one-tenth); 10^{-2} equals 0.1 divided by 10, or 0.01 (one-hundredth). Because *adding* 1 to the exponent easily works out to be the equivalent of multiplying by 10, it is self-consistent that *subtracting* 1 there works out to a division by 10. It is all a matter of placing zeros. Adding another terminal zero is simply to multiply by 10: 100 x 10 equals 1,000. Putting another zero next after the decimal point is to *divide* by 10: 0.01 ÷ 10 equals 0.001. The powers notation makes these operations even clearer.

But what of 10^0? That seems a strange number. However, notice it is equal to 10^1 (10) divided by 10 (or to 10^{-1} multiplied by 10). Although surprising, it is at least logical that 10^0 should be equal to 1.

Because you can make any power of ten ten times larger by adding 1 to its exponent ($10^4 \times 10 = 10^5$), it follows that to multiply by 100 you add 2 to the exponent: $10^3 \times 100 = 10^5$ or 1000 x 100 = 100,000. In general, you can multiply one power of ten by another simply by adding their exponents: $10^6 \times 10^3 = 10^9$. Subtracting the exponents is the equivalent of division: $10^7 \div 10^5 = 10^2$.

All numbers, not only numbers that are exact powers of ten, like 100 or 10,000, can be written with the help of exponential notation. The number 4000 is 4×10^3; 186,000 is 1.86×10^5. This convenient scheme is referred to as scientific notation.

All of this can be extended to basic multipliers other than ten: $2^4 = 2 \times 2 \times 2 \times 2$ (the fourth power of two); $12^2 = 12 \times 12$, and 8^{-1} = one-eighth. (But note that $2^0 = 1$, $12^0 = 1$, and $8^0 = 1$.)

Logarithms arise from extensions of this scheme.

The symbol $\sim$ is mathematicians' shorthand for "approximately" or "about."

It is all very well to say or write 3 x 10^8, but we do not communicate by numerals alone. Many numbers have names, words to designate them. Here follow a few lists of such names, antique and modern, selected from a variety of arithmetical traditions.

Metric number names

The official prefixes used for metric multiples and subdivisions, by powers of 10^3:

atto-	a	10^{-18}
femto-	f	10^{-15}
pico-	p	10^{-12}
nano-	n	10^{-9}
micro-	μ	10^{-6}
milli-	m	10^{-3}
unit		10^0
kilo-	k	10^3
mega-	M	10^6
giga-	G	10^9
tera-	T	10^{12}
peta-	P	10^{15}
exa-	E	10^{18}

Thus kilometer (km) or millimeter (mm). They derive mainly from Greek.

Not quite official:

centi-	10^{-2}
deci-	10^{-1}
deka-	10^1
hecto-	10^2

These assist in fine-tuning nearer the scale of the unit itself.

American number names

units	10^0
tens	10^1
hundreds	10^2
thousands	10^3
millions	10^6
billions	10^9
trillions	10^{12}

The English billion is 10^{12}.

Hindi number names

ek	10^0
das	10^1
san	10^2
hazar	10^3
lakh	10^5
crore	10^7
arahb	10^9
carahb	10^{11}
nie	10^{13}
padham	10^{15}
sankh	10^{17}

Roman number symbols

I	1
V	5
X	10
L	50
C	100
D	500
M	1000

UNITS OF LENGTH

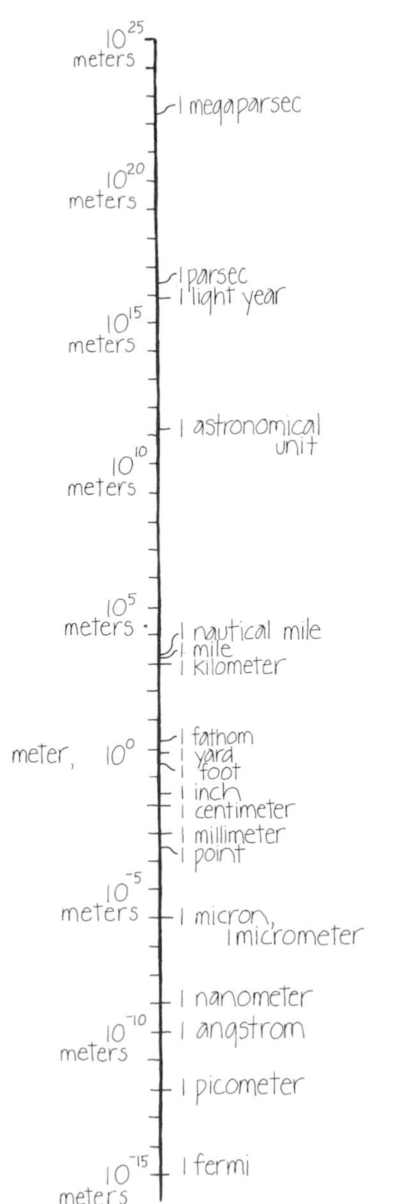

Parisians still consult this public meter displayed near the Palais du Luxembourg *1*. In London it is at Greenwich that you can confirm your yardstick *2*. At *3* is printed a real 10^{-1} meter, or 10 centimeters.

1

2

multiplier:
- × 1000
- × 100
- × 10
- 1
- ÷ 10
- ÷ 100
- ÷ 1000

Grow your own food and build your own house, and no formal units of measurement much interest you; such is the gentle rule of thumb. But commerce has implied agreement on units of measurement. The legal yard has long been displayed for the use of Londoners, and the meter is still open to public comparison on a wall of a Paris building.

The system we call metric is the work of the savants of Revolutionary Paris in the 1790s. Even their determination to celebrate both novelty and reason met limits: Our modern second, minute, and hour remain resolutely nondecimal. That was no oversight—the metric day of ten hours, each of a hundred minutes with a hundred seconds to the minute, was formally adopted. But the scheme met fierce resistance. About the only costly mechanism every middle-class family then proudly owned was a clock or watch, not to be rendered at once useless by any mere claim of consistency! Practice won out over theory.

In much the same way, people who today frequently use units in a particular context are not always persuaded to sacrifice appropriateness to consistency. We list here a few nonstandard units of linear measurement that retain their utility even in these metric days, some even within the sciences.

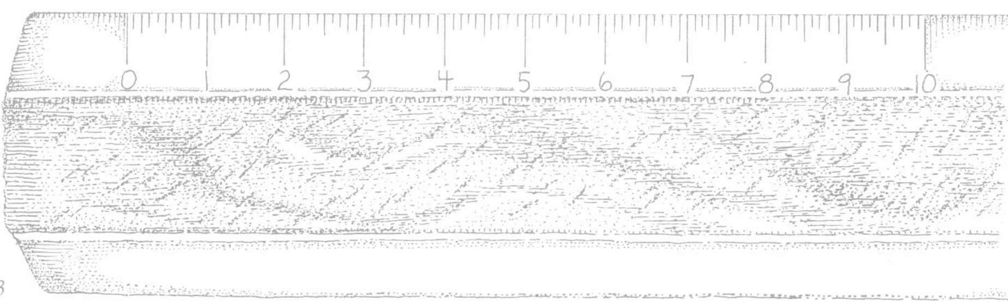

COSMIC DISTANCES

parsec

The word is a coinage from *par*-allax of one *se*cond. The *parsec* is in common use among astronomers because it hints at the surveyor's basic technique of measuring stellar distance by using triangulation. The standard parallax is the apparent shift in direction of a distant object at six-month intervals as the observer moves with the orbiting earth. It is defined so that the radius of the earth's orbit seen from a distance of one parsec spans an angle of one second of arc. The nearest known star to the sun is more than one parsec away.

light-year

This graspable interstellar unit rests on the relationship between cosmic distance and light travel over time.

The speed of light in space is 3.00×10^8 meters per second; in one year light thus moves 9.46×10^{15} meters, which is usually rounded off to 10^{16} meters, especially since only a few cosmic distances are so well known that the roundoff is any real loss of accuracy.

astronomical unit

The mean distance between sun and earth is a fine baseline for surveying the solar system; it is a typical length among orbits. $1 \text{ AU} = 1.50 \times 10^{11}$ meters. *Note*: 1 parsec = 3.26 light-years = 206,300 AU. The interstellar and the solar-system scales plainly differ; intergalactic distances run to megaparsecs.

TERRESTRIAL LENGTHS

miles, leagues, etc.

These are units suited for earth-bound travel, for distances at sea, or for road distances between cities. Nobody ever measured cloth by the mile, or train rides by the parsec.

yards, feet, meters

These rest on human scale, in folklore the length of some good king's arm. They suit well the sizes of rooms, people, trucks, boats. Textiles are yard goods. The meter was defined more universally, but clearly it was meant to supplant the yard and the foot. It was related to the size of the earth: One quadrant of the earth's circumference was defined as exactly 10^7 meters, or 10^4 kilometers. In 1981 the meter is defined with great precision in terms of the wavelength of a specific atomic spectral line. It is "1,650,763.73 wavelengths in vacuum of the radiation corresponding to the transitions between the levels $2p_{10}$ and $5d_5$ of the krypton-86 atom."

inches, centimeters, etc.

The same king's thumb? Human-scale units intended for the smaller artifacts of the hand: paper sizes, furniture, hats, or pies.

line, millimeter, point

Small units for fine work are relatively modern. The seventeenth-century French and English line was a couple of millimeters, and the printer's point measure is about 0.35 mm. Film, watches, and the like are commonly sized by millimeters. The pioneer microscopist Antony van Leeuwenhoek used sand grains as his length comparison, coarse and fine: He counted one hundred of the fine grains to the common inch of his place and time. Smaller measurement units are generally part of modern science, and thus usually metric.

ATOMIC DISTANCES

angstroms, fermis, etc.

Once atoms became the topic of meaningful measurement, new small units of length naturally came into specialized use. The Swedish physicist Anders Ångström a century ago pioneered wavelength measurements of the solar spectrum. He expressed his results in terms of a length unit just 10^{-10} meters long. It has remained in widespread informal use bearing his name: convenient because atoms measure a few angstroms (Å) across. The impulse for such useful jargon words is by no means ended; nuclear particles are often measured in fermis, after the Italian physicist Enrico Fermi. 1 fermi equals 10^{-15} meters.

ANGLES AND TIME

Angles are measured, especially in astronomy, by a nonmetric system that goes back to Babylon. A circle is 360 degrees; 1 degree = 60 minutes of arc; 1 minute = 60 arc-seconds. An arc-second is roughly the smallest angle that the image of a star occupies, smeared as it is by atmospheric motion. This page, viewed from about twenty-five miles away, would appear about one arc-second across.

Time measurement shares the cuneiform usage of powers of sixty. Note that a year of 365.25 days of 24 hours, each of 60 minutes with 60 seconds apiece, amounts to about 3.16×10^7 seconds.

READING THE RAINBOW

Slowly and smoothly turn the dial of the radio. As the marker moves along you are likely to encounter signal after signal, some loud, some weak. It would be no hard task to draw a graph in which the intensity of the sound in the room is plotted against the reading on the dial. There would be a set of peaks, some higher, some lower, with a good deal of space between where only a faint hiss is heard. That graph is a radio spectrum for your vicinity at the time.

Plainly it is not complete: You could turn on the FM set, or the short-wave, or even the TV. All of the signals noted for any of the channels represent incoming energy of a single kind: electromagnetic radiation. Far beyond the dial of any radio you would come at last to the visual region. *Light* is the name we give to electromagnetic radiation in the channels from red to violet, those the eye can detect. An instrument that selects among visible channels is likely to be called a spectrograph, and to record photographically. Signals can be found far beyond the visible, too, and are of no small importance.

The chart presents the whole gamut of electromagnetic radiation through which we gain so much of what we know of the world, all ordered along a single scale. That ordering scale can be labeled in three ways, equally valid. One description of the radiation emphasizes its behavior in time. Like regular water waves, radiation has periodically recurring maxima along its path of travel. At any location along the path, count those maxima for one second; the result is the *frequency* of the radiation, given in cycles per second, a unit now named the hertz (Hz). Another label describes the wave as a moving pattern in space: The distance at any instant between one crest and the spatially adjoining crest is the *wavelength*, measured in meters or in multiples of the meter. Finally, for every kind of radiation a maximum energy can be transferred in a single atomic interaction; that is called the *quantum energy*, measured in electron volts. (One electron volt is the energy of an electron that can just make its way against a certain standardized electrical repulsion.)

At the radio end of the spectrum the label usually used is frequency, for radio techniques make time variations discernible, something almost impossible at the other end of the spectrum. In the middle range, the wavelength is much used, for between the millimeter band and the x rays, techniques are quite available for exhibiting spatial variation. At the high-energy gamma-ray end, energy is the most used label; at this end it is easy to measure the sizeable energy transfers, while the intrinsic variations in space and time are alike inaccessible.

All these radiations have common physical qualities. Their speed in empty space is the same: It is the ultimate speed, supersonic by six powers of ten. All radiations are emitted and absorbed only by the motions of electric charges (and by related magnetic effects). Therefore as a class they are called electromagnetic. A sound wave of frequency one thousand cycles per second (1 kilohertz) is not at all the same phenomenon as an electromagnetic wave of identical frequency, though the electronic means to convert one to the other are commonplace. The electromagnetic wave is a pattern of varying electric and magnetic fields persisting even in empty space; the sound wave is a pattern of pressure and motion maintained in the air or another material medium, absent in vacuum.

Nearly all electromagnetic radiation, from the lowest frequencies up to quantum energies in the range of 100 thousand electron volts, is the product of energy transfer by moving electrons, either singly in atoms or in myriads within a current-carrying wire. Above that energy limit the radiation is often not electronic but nuclear in origin, coming from similar motions of charged protons. At still higher energies, over 100 million electron volts, the energy transfers that can give rise to or absorb the radiation often involve subnuclear motions of quarks or of other novel charged particles.

THE ANALYSIS OF SPECTRA

In the living room one could turn the radio dial and count up and order the radio transmitters in operation nearby. That would be a local radio spectrum. In the analogous way one can record spectra in the optical region. Those spectra give a chemical analysis of the glowing matter that is the source of the light, whether it be on the lab bench half a meter away, or on the surface of a distant star. The transmitters for specific visible frequencies turned out to be particular atoms and molecules. The channel frequency of a radio station is determined by the construction of key elements of the radio transmitter; channel frequencies for atoms are similarly intrinsic to atomic structure. Long before we had much idea of how atoms are made, we could treat spectral patterns observed in the lab as a kind of atomic signature. Once the same pattern of optical channel locations were seen in starlight spectra as in lab spectra, it was easy to infer the elements present.

Two kinds of spectra are shown in the pages at 10^{18} meters. The range of color seen from the Ring Nebula resembles the result of our radio-dial study: intense peaks of color on a dark background. The other kind is the spectrum of Arcturus—dark interruptions in a general glow. There an intervening layer of gas has imposed its own signature by its preferential interactions, absorbing the starlight passing through it. The glow arises in the deeper and hotter region below, where the atoms that make the light are so dense and the layer so thick that the radiation coming from below loses, through repeated interactions, its sharpness of frequency definition.

More remarkable still is the kind of result shown in the figure on page 14. It is a spectrum seen from atomic hydrogen in a laboratory electrical discharge. No need here to analyze for elements; the atomic composition is established. But now the regularity and simplicity of this spectral signature of a single atom—the simplest—are clear. Bohr could use the invisible order, the numerical relationships among the frequencies of those characteristic spectral lines of hydrogen, as a kind of analogue to Kepler's laws (compare p. 9). Quantum motion is not directly visible like the motions of the planets, but its spectral signature led to the atom's structure, the invisible order among electrons bound by the nucleus. It goes without saying that gamma-ray spectra emitted by nuclei, and by the quark-clouds we know as free protons and neutrons and the rest, are fine raw material for structural probes of the ultramicroscopic world. Spectra reveal the persistent quantum states within every radiating system.

Finally, if a spectral pattern is recognizable in a celestial body, say by the correct ratios of several frequencies, it may happen that individual peaks do not coincide in frequency with laboratory standards. All of them are, for instance, shifted to the red by one part in a thousand. The interpretation is easy: A fast-receding source sends to a detector a radiation signal in which as time goes by the distance of travel grows steadily greater. That has the effect of lengthening the time between received crests of the waves, and so shifting the frequency in proportion. An approaching source is seen blue-shifted. The speed of any source relative to the detector can be measured at once; no need to wait out position change over centuries. This shift, the Döppler shift as it is called after its nineteenth-century Austrian discoverer, works for any radiation band, provided an emission frequency pattern can be recognized. It is a touchstone for all sorts of astronomical studies; and it is the only way to distinguish the rapid recession of distant galaxies—the famous cosmological red shift—believed to be the sign of a steadily diluting universe at large scale.

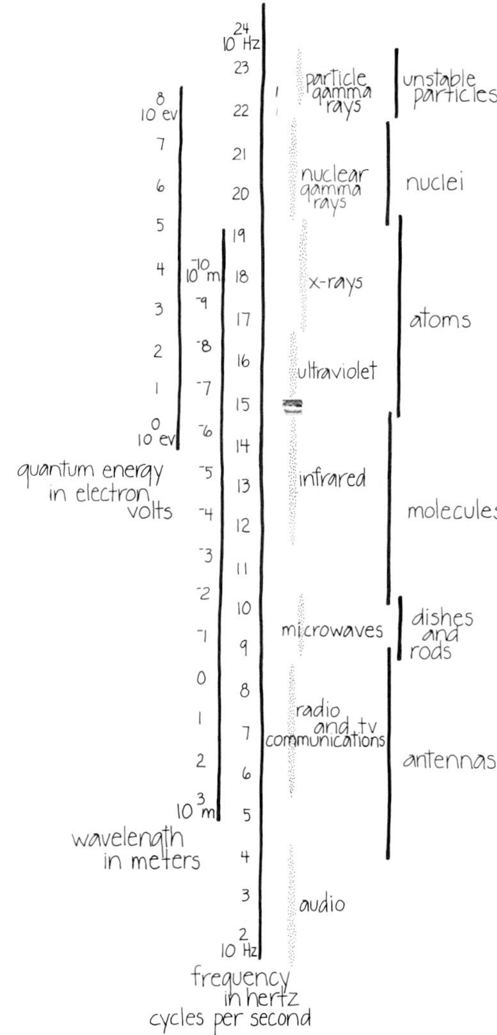

Electromagnetic radiation with a wavelength of one micron, 10^{-6} meters, lies just redward of what the eye can see. (That range is called the near infrared.) It has a frequency of 3.00×10^{14} hertz and a quantum energy of 1.24 electron volts. One example will fix the relation of the three scales; for frequency is strictly proportional to quantum energy and inversely proportional to wavelength. (1 eV $= 1.60 \times 10^{-19}$ joules, the everyday metric unit of energy; a single light quantum has a tiny energy, appropriate to a single atom.)

Our grasp of the universe has grown along with the instruments that in one way or another extend the work of the senses, in particular the eye.The seventeenth century brought us two optical devices whose names have by now passed into everyday use: the telescope and the microscope. A book devoted to the visual exploration of scale can hardly fail to celebrate these two in particular, in both older and newer forms.

Why should we trust the knowledge brought us by such unnatural means? Indeed, we ought not to trust it; each instrumental result should be viewed skeptically until the performance of the device has been deeply tested. But exactly the same caution applies to our inborn instruments, the eye and the other senses that serve human perception. They all have their limitations, their illusions, their defects. Most of us learned in childhood much of the nature of instrumental signals from eye and ear; that interpretation is part of individual human development.

Accepting what our senses and our instruments tell us does not necessarily imply a knowledge of the principles of design and function of the instrument itself. The eye, like the telescope, can be viewed as a black box, a piece of sealed equipment whose relevant properties can be judged best by use, by the painstaking tests of consistency and meaning. An afterimage is not a violet spot floating along the wall; we learn that from early experience and not from studying the recovery of exposed retinal pigments. Of course scientists hope one day to understand the whole chain of perceptual events. When the critics quite rightly complained that the lenses of Galileo's telescope might be producing mere illusions, as with the curved mirrors just then rather popular, they could be answered best by tests on the instrument itself. Galileo observed that it would be a subtle illusion indeed that showed four moons moving in repetitive patterns around Jupiter, but none near Venus or Mars. That has the ring of the authentic beginning of modern science; it leads to a view of knowledge just as cogent for everyday witness as for the learned judgments of the laboratory.

Optical telescope and microscope are extensions and modifications of the eye, which also depends on a lens. The lens of the eye projects a small image of the scene on the retina. The telescope works by using a lens to produce a nearby image of a distant object, which can be examined close at hand to yield a magnified view. The microscope has instead a lens that can image an object very nearby at a size suited to detailed examination.

Mirrors can be used to focus light in much the same way as lenses. To collect enough energy for viewing distant weak sources requires a large collecting area; telescopes (and their mirrors) grew large. Microscopes, on the other hand, developed in image-forming quality and in mechanical nicety and stability. Their extension beyond visible light came once our technology allowed detection of radiations invisible to the evolved eye.

1

2

3

This instrument site *1* was built around 1584 by Tycho Brahe, a nobleman of Denmark, adjoining his great headquarters observatory on the island of Ven, in the sound between Denmark and Sweden. Masonry pits held permanent instruments for the accurate measurement of directions in the sky. Careful sighting along elegant scales in heavy and ornate frames of steel, wood, and brass spoke the last word in naked-eye precision, in a tradition two thousand years old, to be outdone within a few generations by glass lenses. Yet the data scrupulously taken here became the raw material for Kepler's laws.

These two telescopes *2* were among many made by Galileo himself, working in his own shop along with an expert mechanician and glassworker. Their first good instrument was made late in 1609, on the basis of a report that "a certain Dutchman" had made a spyglass through which distant objects were seen as if nearby. There seems no doubt that Galileo developed both the telescope and how to use it, quickly raising the device from marginal utility to become the foundation of a new cosmos. Spyglasses of low power were not uncommon by 1609, for public sale in Paris as in the Netherlands, but most of them were optically seriously inferior to these. Tycho's precision was soon left one or two powers of ten behind, and celestial novelties poured in as well in the new look at the small and the faint.

John Herschel took the telescope *3* in 1834 to a countryside site six miles out of Cape Town. He used it for four years in a systematic survey of the southern skies, studied for the first time with a substantial instrument. His goal was to commemorate and extend the lifelong work of his famous father William (and his aunt Caroline) in mapping English skies. Its mirror was 18 inches across; the simple wooden mounting was in the family style. Three interchangeable mirrors were on hand: One had been made by William, another by father and son together, and the third was John's own. There was no photography as yet, but sharp eyes and the big mirror.

4

The evocative period photo from 1916 shows the base of the tube of the 100-inch reflector *4* with its escort starting out for the summit of Mt. Wilson, just north of Pasadena, California. Such an instrument is plainly a precision piece of heavy engineering. It was this photographic telescope that made possible the first credible scaling of our Galaxy and the final demonstration that we live in an island universe of external galaxies.

5

Los Angeles grew so big and bright that Mt. Wilson telescopes no longer saw a sky dark enough to study faint objects (except during the brownouts of World War II). A still greater telescope was planned in the 1930s, delayed by the war, and finally built on Palomar Mountain between Los Angeles and San Diego. This wonderful drawing made during the long planning stage shows a cross section of the proposed dome and instrument *5*; the small figure of an observer can just be made out at the highest point of the big vertical skeleton tube. The mirror is 200 inches (5.08 meters) in diameter; finished in 1948, this telescope is still the second largest in the world, though in the Caucasus Soviet astronomers have recently begun use of a telescope of full 6 meters aperture. The 200-inch can claim to its credit, for example, the discovery of quasars. Photography is now beginning to give way at the telescope to newer electronic imaging devices.

6

Radio astronomy wavelengths are measured in tenths of meters, whereas visible ones are around a half a micron. Radio telescope mirrors therefore need a very large physical diameter to match the relative scale of their optical counterparts. This mirror in the green hills near Arecibo, Puerto Rico is the largest radio dish *6* in the world, 300 meters in diameter. Its surface of thin perforated aluminum plate is held true to a fraction of a radio wavelength, as an optical mirror is true to a fraction of a micron. The central structure high above (it has the size and appearance of a railroad bridge; look for its shadow on the bowl below) supports a moving antenna at the focus of the big mirror. The fixed dish can, in this way, follow a moving source in the sky to record weak natural signals for a couple of hours as the earth turns.

7 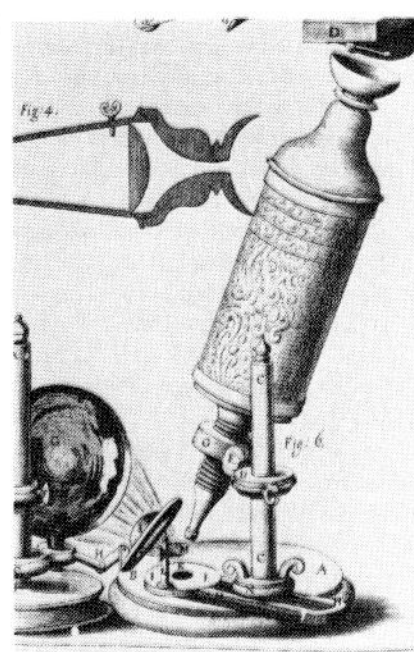

8 

In 1665 the versatile Robert Hooke, brilliant pioneer of experimental science, published a book called *Micrographia*. It is a masterpiece, and met a lasting popular success through its lively text and its wonderful large illustrations of the newly found invisible world at a magnification from fifty- to a hundredfold: needles, soot, fleas, linen, mold, cork, feathers. . . . The drawing 7 is made from the frontispiece of the book; it shows Hooke's own microscope with its accessories. Mr. Samuel Pepys bought the volume at once and sat up until two in the morning to read it: "The most ingenious book I ever read in my life," he noted.

Optical microscopy has grown in deftness for centuries. By now the small world can be studied in detail from millimeter scale down to a fraction of a micron; the limit is set by the wavelength of visible light. Optical microscopes can examine living organisms—wonderful cinema—for the watery medium these require is transparent. Detail can be revealed by clever use not only of real color differences, but of the ones that arise by optical exploitation of the density differences of transparent materials. Using reactive dyestuffs to mark structures by taking advantage of their chemical differences is another powerful technique. These microscopes work best on thin layers; the depth of sharp focus is small.

By the 1930s the control of electron beams in a vacuum had matured. Now the transmission electron microscope (TEM), which forms an image by using magnetic analogues to lenses, can show remarkable cross-sectional detail down almost to the atomic level of a few angstroms. The scanning electron microscope (SEM) employs instead a fine pencil-beam of electrons moving in TV fashion to paint out an image over time, using the electrons that scatter off the successive target points.

The medium in which electrons travel undisturbed is the vacuum; but a vacuum cannot support life. So under electron microscopes we view not living but dead structures. The transmission electron microscope, like the light microscope, requires thin-layer specimens; the preparation of such sections has become a virtuoso technique. The transmission electron microscope can work well for structures from the micron range down to the order of 10^{-9} meters; with special instruments and techniques we can glimpse individual atoms.

The strength of the scanning electron microscope is its wonderful depth of focus; it can image structures from millimeter scale down to 10^{-8} meters, the limit set by the submicron fineness of its scanning beam.

The big accelerators that bring beams of protons up to very high energies are like microscopes, if only in function, for these ultramicroscopes carry out probes of spatial structure at the smallest scales. The result is much more indirect than any ordinary image; here the mechanics of the familiar world no longer hold, and a subtler indirect analysis is the rule. This complex 8 is the Fermi Laboratory. The long tangent line through which the beam of fast protons exits for study along with its secondary products is evident. But the dominating feature is the big main ring of underground magnet sections, visible at the surface as a circle two kilometers in diameter, much larger than the radio dish at Arecibo. This laboratory is located just off the western edge of our Chicago picture, 10^5 meters.

CHRONOLOGY

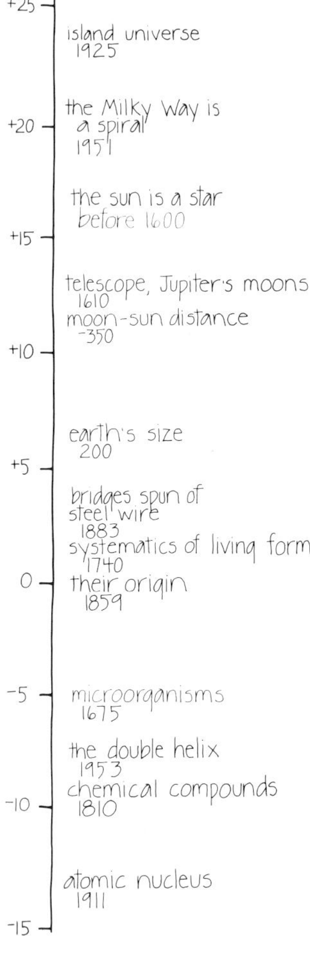

This list is a selection of the dates and authors of discoveries, inventions, early developments, and first persuasive formulations of concepts that have built the present understanding of the world. The list is not ordered by date, but roughly by the physical scale to which each finding is related. It shows when we came to understand each order of magnitude.

We know that a simple listing is somewhat misleading; the history of these ideas and experiences is too complicated to describe fully in so summary a form. Even the dates are a little uncertain; sometimes they refer to discovery, sometimes to publication. Most of these discoveries have had their forerunners and their rivals. We have tried to point out the influential beginnings. Some famous names are included, partly to place the great works in time. The list favors events with strong effect on the visual model and on the underlying evidence.

10^{25} meters

Universal background radiation found	1965	Penzias & Wilson
Quasars recognized	1963	Schmidt
Clustered galaxies share a common motion	1917	Slipher
"Island universe"; galaxies are far-away Milky Ways:		
—speculation	1755	Kant
—verification	1925	Hubble
Nuclei of some galaxies are active	1943	Seyfert
Spiral forms seen in nebula	1850	Lord Rosse
Galaxies contain populations of old stars and new	1944	Baade
New channels open for astronomy:		
—radio (Milky Way)	1933	Jansky
—first distant radio galaxy	1949	Bolton, Stanley & Slee
—x rays (beyond the sun)	1962	Rossi, Giacconi, Gursky & Paolini
—infrared (in deep sky)	1968	Becklin & Neugebauer
—electromagnetic spectrum	1867	Maxwell
	1880	Hertz
Magellanic Clouds made known to Europeans	1516	Corsali
Andromeda Nebula seen:		
—unaided eye	970	Al-Sufi
—telescope	1612	Marius

10^{20} meters

Our galaxy, the Milky Way:		
—its size	1918	Shapley
—its spiral form	1951	Morgan, Sharpless & Osterbrock
The Milky Way is a collection of stars	1610	Galileo
Dark nebulae:		
—seen by eye	prehistoric	
—shown made by dust	1923	Wolf
Bright nebulae seen:		
—in telescope	1610	Peiresc
—to have gaseous content	1864	Huggins
—to be illuminated by bright stars	1922	Hubble
Supernovae recognized as a class	1934	Baade, Zwicky
The sun is a star, and all the stars are suns:	Around 1600	
—first quantitative results	1684	Huygens
A double star follows the laws of gravity	1830	Savary
The stars (like the sun) are of familiar chemical elements	1863	Huggins
Star-spectra photographs	1872	Henry Draper
Star distances measured by triangulation	1838	Bessel

10^{15} meters

Distant comet cloud around solar system	1950	Oort
Comets are beyond the atmosphere	1577	Brahe
Comets follow Newton's laws, some do return	1705	Halley
The earth as a sphere	About −500	The school of Pythagoras
Circumnavigation of earth	1520	Magellan
First fully triangulated national map	Begun 1744	Cassini de Thury
Longitude at sea: the chronometer	1761	Harrison
Artificial earth satellite:		
—theory	1687	Newton
—realized	1957	USSR
The weather as a general pattern	1686 to the 1740s	Halley, Hadley & D'Alembert
The atmosphere estimated in height and weight	1640s	Pascal, Torricelli
Hurricanes described	1492	Columbus
America revealed to Europe	After 1492	Columbus
The role of glaciation in the past	1830s	Bernhardi, Hitchcock & Agassiz
The Great Lakes as a glacial legacy	1880s	U.S. Geological Survey
Systematic earth photos from orbit	1972	Landsat, NASA

10^5 meters

Everest shown the highest mountain	1852	India Survey
The deep Pacific ocean trenches sounded	1875	U.S.S. *Tuscarora*
Domestication of wheat	About −8000	Uplands from Palestine to the Zagros Mountains
First cities	About −4000	Euphrates Valley
Bridge cables spun of steel wire	1883	John & Washington Roebling
First use of fire	Before -10^6	Early humans along the African rift
First considerable steel-framed building, Chicago	1890	Burnham & Root

10^0 meters

Systematics of the species of animals, vegetables, minerals	1740	Linnaeus
Origin of species explained	1859	Darwin
Logarithms	1594	Napier
"Instantaneous" photography (with fast gelatine dry plates)	1880	Burgess & Kenett, Bennett
Microcomputer (one landmark in chip development from 1960s)	1972	Intel Corp.
Circulation of the blood	1628	Harvey
Optical microscope	About 1610	Developed by several, including Galileo
The capillary connection	1661	Malpighi
The red cells of the blood seen	1674	Van Leeuwenhoek
Cells described as a general feature of life	1839	Schwann

10^{-5} meters

The cell nucleus seen	1831	Brown
DNA:		
—first isolated	1864	Miescher
—can carry heritable information	1944	Avery & McLeod
—the double helix	1953	Watson & Crick
Bacteria and protozoa seen under microscope	1675	Van Leeuwenhoek
Bacteria grown in pure culture	1881	Koch
Lymphocytes form part of the human immune system	1882	Pasteur, Metchnikoff
Lenses for electron beams	1925	H. Busch
Scanning electron microscope	1938 1953	Von Ardenne McMullan & Oatley
Transmission electron microscope	1931	Knoll & Ruska
An enzyme isolated as a crystalline protein	1926, 1930	Sumner, Northrup
Electricity is particulate, and chemical forces are electrical	1881	Helmholtz
The electron is found, weighed relative to hydrogen (and named)	1897	J. J. Thomson
The hydrogen atom recognized as the lightest of atoms; atomic weights	1858	Cannizzaro
Chemical compounds and their atomic nature	1810	Dalton
Specific bonds for specific atoms	1852	Frankland
Ring molecules visualized in space	1865	Kekulé
Modeling of three-dimensional molecular forms	1874	Van't Hoff & Le Bel

10^{-10} meters

Molecular model of solid, liquid, gas	1870s	Maxwell, van der Waals
The periodic system:		
—chemically based	1869	Mendeleef
—first physical theory	1922	Bohr
The Bohr theory of the atom: hydrogen and its spectra, early approach to quantum motion	1913	Bohr
The true quantum mechanics	1925	Heisenberg, Dirac, Schrödinger
Quantum theory of chemical bonds	1930s	Heitler & London, Pauling
X rays	1896	Roentgen
X rays emitted from inner shells of the atoms: the atomic number	1914	Moseley
X-ray analysis of crystalline arrays of atoms	1912	von Laue
The nucleus of the atom found and measured	1911	Rutherford
Regularities of optical atomic spectra	1884	Balmer
Hydrogen ions as protons	1912	J. J. Thomson
Isotopes named	1913	Soddy
The neutron	1932	Chadwick
The positron:		
—predicted	1928	Dirac
—found	1932	Anderson
The nucleus is built of neutrons and protons	1932	Heisenberg
The neutrino:		
—predicted	1930 1934	Pauli Fermi
—directly detected	1956	Reines & Cowen
The π-meson:		
—predicted	1935	Yukawa
—found	1946	Powell
Proton accelerators	1930 1946	Lawrence MacMillan, Veksler
Plethora of transient particles, the mesons and hyperons	1960s–1970s	

10^{-16} meters

Quarks grow to credibility	Late 1970s

SOURCES AND NOTES

The data and the images that make up this book arise out of the body of science and technology itself. But that alone does not bring them together. In this section we want to thank explicitly the many people from many places who generously made available their expert work for use here. We list the origin of every image in some detail. Then we often take the opportunity to amplify a little, sometimes on the science and technology of image-making itself, sometimes on the background and implications of what you see.

page 1
The uncaptioned figure that shows an eye caught by the sight of the two spheres of heaven and earth is a woodcut from *Cosmographicus liber*, by Peter Apian, editor, Gemma Frisius, Antwerp, 1533. The photo is from the Huntington Library of San Marino, California.

page 2
The miniature terrestrial globe in a case, which is itself a concave sky globe showing constellation figures, was a popular elegance of the eighteenth century. This one was made in London about 1750; it claims to record the "New Discoveries by Dr. Halley." The globe and its photo are from the Antique Instrument Collection at The Adler Planetarium, Chicago.

page 5
The theoretical cosmographer Oronce Finé himself drew this illustration of the spheres of the elements for the *Livre singulier et utile, touchant l'art et practique de geometrie*, by Charles de Bouelles, Paris, 1542. The illustration foreshadows the pyramidal view of our journey. Photo from the Huntington Library.

page 6
The satellite Phobos, of Mars, is shown (1971) in a Jet Propulsion Lab photo: NASA Mariner 9, PHOBOS JPL/ORBIT 34.

page 7
The satellite Dione of Saturn shown in a photo by Voyager 1 from 380,000 km away, taken November 1980. JPL photo P-23215.

We venture a little farther into the several physical forces. Gravitation dominates the large-scale world, as Phobos suggests. From the scale of 10^6 meters and upwards, it is gravity which binds almost all structures. They are, in general, kept by internal motions from final collapse under that insistent attraction. Between the scale of 10^6 meters and that of 10^{-14} meters is the domain of cohesive matter, from mountains to molecules and atoms. So long as we deal with matter beyond the minute reach of the strong nuclear forces, a complex set of balances is struck between electrical attraction and repulsion. These forces control all structure, with the participation of motion—both the random motions we know as thermal and that which we characterize as quantum motion. Those motions play their parts especially well on the molecular and atomic stages. Below the level of the innermost electrons, at the 10^{-14}-meter scale and below, we may enter under the sway of the nuclear forces themselves. Mediated by meson exchange, they are powerful though their range is so small. Within the nucleus the specific nuclear attractions and repulsions need to settle their affairs both with the quantum motion and with the mutual electrical repulsion of the positively charged protons. The results fix the details of nuclear structure. Farther below that, at 10^{-15} meters and less within the proton, we come to the binding of quarks by the novel forces called chromodynamic, a complex mediation by a set of particles named *gluons*. With them we near the end of current knowledge.

page 8
A contemporary drawing done over an air photo of the Chicago World's Fair. Photo taken March 2, 1933 by the Chicago Aerial Survey Co., Des Plaines, Illinois, adapted by artist William Macy. From the files of the Chicago Historical Society: ICHi-16222.

page 9
Kepler published this relation in 1619. We used modern data. Nowadays this law of Kepler defines an important natural resource. Objects orbiting the earth will obey a similar proportion between their orbital times and their distances from earth. The moon orbits earth at a distance of about 240,000 miles in around 29 days; a satellite placed at a distance of about 26,000 miles from the center of the earth turns out by Kepler's relation to have an orbital time of just one day. If the artificial satellite not only is placed at the right distance but is arranged to circle in the plane of the earth's equator, the satellite will appear to remain at a fixed point in the sky, neither rising nor setting for a watcher—or an antenna—on the ground below.

That result of Kepler's describes the geosynchronous orbit. It is so practical that scores of weather and communications satellites dot that big orbital circle in space today, the makings of an artificial ring around earth like Saturn's own.

page 11
This x-ray diffraction photo with its remarkably fine detail is the work of a 1979 collaboration between a group at MIT and one at the University of Leiden. Photo supplied by Andrew H. J. Wang of MIT.

page 12
A tiny additional discrepancy in this sort of motion long known for the orbit of Mercury was explained in 1916 by Einstein's new gravitational theory, in its first triumph.

page 14
The hydrogen spectrum showing the lines of the Balmer series is after the drawing by Roger Hayward in Linus Pauling's *General Chemistry*.

page 15
The mica hand, ten inches long, is in the collections of the Marshall Field Museum of Natural History, Chicago.

10^{25} meters

The overall regular expansion of the original gas in the universe led to the mutual recession of galaxies which grows with their separation. Two galaxies that lie at opposite edges of this largest square drift apart at roughly 20,000 km/sec, not much less than one-tenth the speed of light.

It is this motion which shows up in the spectra of distant galaxies and quasars. To go beyond 10^{25} meters to the next tenfold step would squarely confront relativistic physics.

The high speed of the very distant galaxy is judged from the spectral shift of a few recognizable atomic lines. They appear redward in wavelength position by a factor of almost two. Lines seen from this object in the red are, upon emission, intrinsically blue; the lines are identified by matching their wavelength ratios. Enlargement from a direct (red) plate taken about 1977 by H. Spinrad and H. E. Smith at the 4-meter telescope of The Kitt Peak National Observatory, Tucson, Arizona.

2 The regular recession of distant galaxies is used to assess the distance of otherwise unfamiliar objects so far away that their overall motion outweighs the gravitational effects of neighbors. The quasars are the best examples. Once the recession speed is one percent that of light, the method is secure—if our understanding is correct. Most quasars show shifts by more than a factor of two, always to the red, because expansion carries them all outward from us (as we would fly outward seen from any one of them). Source 147 in the Third Cambridge University radio source list is a quasar. The photo of 3C147 was made in the late sixties with the 200-inch Hale telescope at Palomar Observatory, Pasadena, California.

3 A faint cluster of galaxies in Pisces, from the 200-inch telescope. Palomar Observatory photograph.

10^{24} meters

1 The center region of the Virgo Cluster, showing M84 and a number of other bright galaxies, from The Kitt Peak National Observatory 4-meter telescope.

2 M84, or NGC 4374, a bright elliptical galaxy of the Virgo Cluster, in a photo from The Kitt Peak National Observatory 4-meter telescope.

3 The galaxy pair NGC 4038/39 emerging from a lengthy collision, in a photo by Francois Schweizer with the 4-meter telescope of the Cerro Tololo Inter-American Observatory in Chile. Their computer simulation of this event is shown by A. Toomre and J. Toomre in *Scientific American* December 1973. We know a number of colliding galaxies. Though the density of matter is much greater inside the galaxies than outside them, we see no collisions between stars. Relative to their own sizes, stars are far more widely separated than are the galaxies. Star collisions would be short-lasting events compared to these galaxy encounters.

We have two ways to estimate the total substance within galaxies and their clusters. One way is to count up the stars and all other radiating sources, then to reckon their total mass using our local understanding of galaxies, gas, and stars. The other is to work out the gravitational pulls from their connection to observed speeds and forms of the big structures.

It has become fairly sure that there is an important discrepancy between the two estimates. Much of the gravitating matter in the universe—perhaps 90 percent of it—seems to be quite invisible. What can it be? The most conservative view is that most galaxies bear big haloes of dull reddish dwarf stars, too faint to see. Already much of the mass of a galaxy is known to reside in such stars; maybe there are ten times more of them? The most novel proposal, jazzy and exciting, is that neutrinos—individual subnuclear particles we have good reason to expect in space but cannot now detect at all—provide the invisible mass, their own mass. This is radical Copernicanism; for if it were true, not only is our earth no center of the universe, but the very matter of our stars, our earth, and ourselves would be a sort of impurity in a universe whose major present substance is the orbiting invisible neutrinos! For now we do not know.

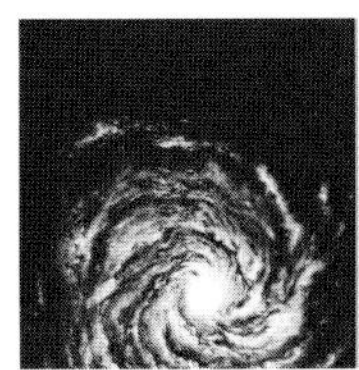

10^{23} meters

1 The two photos of the galaxy
2 pair NGC 5426/27 were made with similar long exposure times at the 4-meter Cerro Tololo telescope in 1977, by Vera Rubin, Carnegie Institution of Washington, D.C. The wide-band photo used blue-sensitive emulsion; the other, a narrow filter for the red Hα line of hydrogen.

3 The peculiar galaxy NGC 5128, not far beyond the Local Group, in a 200-inch photo, Palomar Observatory.

4 Redrawn from a radio map of NGC 5128 made about 1965 at 408 megaHertz with the 210-foot radio telescope of the Commonwealth Scientific and Industrial Research Organization at Parkes, NSW, Australia, by Cooper, Price, and Cole, *Australian Journal of Physics*, 1965.

10^{22} meters

1 The Large and Small Magellanic Clouds, in a wide-angle photo made with the 3-inch Ross-Tessar camera at the Boyden Station of Harvard College Observatory in Bloemfontein, South Africa, in 1934. This was a three-hour exposure on blue-sensitive emulsion. Courtesy Harvard College Observatory.

2 The first account of the south polar sky was published by Andreas Corsali in Florence in 1516. It was translated into English and published by Richard Eden in London, 1555, in *The Decades of the New World or West India*. The figure reproduces Eden's redrawing of the Clouds of Magellan. Courtesy the Library of Congress.

3 This night-sky photo of the Andromeda galaxy as it appears to the stargazer is the work of W. K. Hartmann of Tucson, Arizona. He took this wide shot as a twenty-minute time exposure with a 35-mm camera. It is published as one illustration in his fine textbook of astronomy (see our Resources). You might be able to pick out the lazy 'W' of Cassiopeia on the upper edge of the Milky Way band.

4 The same Andromeda galaxy seen with the magnification and light grasp of the old 36-inch Crossley reflector at Lick Observatory.

10^{21} meters

Dr. James Wray, Department of Astronomy, University of Texas, has developed a technique especially for the color photography of galaxies. The image through the big telescope is displayed by a television-like device in the color of a monochrome TV screen. In this way he takes three separate negatives of the object, each as viewed through an appropriate color filter. He superimposes the three images, adding to each the single needed color by a dye transfer process that allows detailed control of the result. (A forerunner of his procedure is known as Technicolor.) In this way he has collected many galaxy images in comparable valid color, though an eye at the telescope could not see color in so faint an image. We are grateful for his permission to reproduce a number of his images.

1 M87

2 NGC 2841

3 NGC 3745

4 NGC 488

5 M88

6 NGC 891

7 NGC 4449

8 M104

9 M51

Charles Messier published in 1784 a list of fixed nonstellar objects that might confuse a comet-seeker like himself who used a 3- or 4-inch telescope. The Messier catalogue includes the brightest, best-seen objects of that kind, at least in the skies visible from Paris. There are only a hundred-odd Messier numbers, so they are often used. NGC stands for *New General Catalogue*, an Edwardian compilation of about 13,000 nonstellar objects.

10^{20} meters

1 Tycho's map of the remarkable new star appeared in his first book, *De nova . . . stella*, published in Copenhagen in 1573, when its author was twenty-seven. The event probably impelled Tycho to continue in astronomy after a year's diversion into chemical experimentation. Photo by Charles Eames.

2 X-ray astronomy is only two decades old. It must be done from above the bulk of the atmosphere; most of its data have come from instruments in earth orbit. The latest of these was the Einstein Observatory (HEAO-2), launched by NASA in 1979 and active until 1981. This figure is a photon-by-photon reconstruction, to generate a computer map of the Tycho remnant in x rays of a few thousand electron volts photon energy. It is the work of Paul Gorenstein, Harvard–Smithsonian Center for Astrophysics, Cambridge, Massachusetts.

3 This is a radio map of the same Tycho remnant at the same scale, published in 1975, taken with the Westerbork radio telescope in the Netherlands by Duin and Strom.

4 The obscure optical remnant of Tycho's star is shown in a red photo made some years ago at the 200-inch telescope of the Hale Observatories, by S. van den Bergh.

5 The Crab Nebula is shown here in the red band on a plate of the 200-inch telescope. Courtesy Palomar Observatory.

6 7 This striking color pair taken one year apart shows the galaxy M100 with and without its supernova of spring 1979. This is the work of a leading amateur astronomer, whose techniques with cooled film get remarkable results from a 14-inch reflector. He is Ben Mayer of Los Angeles, a professional designer. The work was done at his Problicom Observatory, in the mountains northwest of the city.

The citation is the opening sentence of *View from a Distant Star*, by Harlow Shapley, Basic Books, New York, 1963.

10^{19} meters

1 In Sagittarius, right on the Milky Way plane, this is one of the best known of bright nebulae. It is about 5000 light-years away. Its three-part form has given it the popular name Trifid to add to M20 and NGC 6514. The photo was made with the 120-inch reflector at Lick Observatory.

2 The dusty dark Coal Sack was so named by English observers in South Africa in the nineteenth century; but dark constellations that include it are traditional among southern peoples who know their skies. Near the Southern Cross, it is also in the plane of the Galaxy, about five or six hundred light years away. The photo is another from Harvard's Boyden Station near Bloemfontein, a wide-angle shot with the 3-inch camera, taken in 1945. Courtesy Harvard College Observatory.

3 This photo of the Milky Way is another of W. K. Hartmann's own shots, with a 35-mm camera and a wide-angle lens. It spans about fifty degrees along the Milky Way. From Hartmann's text. (See "Resources.")

4 The photo is made with the 4-meter telescope of the Cerro Tololo Inter-American Observatory and cropped especially to compare the edgewise galaxy NGC 55 with the edgewise view of our own Milky Way.

10^{18} meters

1 This star field is centered on the Ring Nebula in Lyra, NGC 6720. That planetary nebula is a popular object for the small telescope; it is nearly two thousand light-years away, and it has been expanding at the modest speed of 20 km/sec for some ten thousand years. This rather early color photo was made through an objective prism on the Burrell Schmidt-type wide-field telescope of the Warner and Swasey Observatory, Case Institute of Technology and Western Reserve University, Cleveland, Ohio. Red hydrogen emission and several bluish emissions of atomic hydrogen and oxygen are seen. We thank Peter Pesch, Director, for this unusual photo.

2 The spectrum of Arcturus shown in detail. The portion reproduced extends only from 4200 angstroms to 4300 angstroms wavelength: It is all within the violet, yet shows many dark atomic absorption lines. Arcturus is a giant star, a hundred times more luminous than the sun, perhaps twenty times larger in diameter, and only around 4000 degrees Kelvin in surface temperature (compared to the sun's 5800 K). It is relatively puffy and cool; indeed, it looks orange to the eye. Mount Wilson and Las Campanas Observatories, Carnegie Institution of Washington.

10^{17} meters

1 This is a small star field near the galactic plane in Lacerta, a fraction of a degree across. On one portion those stars are marked that are known to have a spectral type about the same as that of our sun. Most of them—though they may be multiple stars—must be as luminous, stable, and enduring as is the sun. (About all we know of them is what is marked on this image.) We can estimate from such samples how many sunlike stars the Galaxy holds: plenty! The marked section is very small, well under one-millionth of the whole sky. We do not yet know of any other star with planetary companions like the sun.

That there are very many enduring sunlike stars in the Galaxy is a near certainty. It is an old conjecture that somewhere near such a star there might live evolved counterparts to ourselves. For about twenty years it has been realized that radio signals from sources no different from those we now operate on earth might be detected across interstellar distances, whether used as deliberate beacons or incidental to other ends. No systematic search for such signals has yet been organized; modest pioneer surveys have found no signals.

The photo was prepared by Jesse Greenstein, Hale Observatories, and is reproduced from *SETI: The Search for Extraterrestrial Intelligence*, NASA SP-419, Billingham, Morrison, and Wolfe, eds., Dover, New York, 1979.

2 The double star is plain. It is 61 Cygni, famous because it was the first star to have its distance measured by parallax, the feat of F. W. Bessel in 1838. The two stars move very slowly, passing around their mutual orbit in about eight hundred years. They have drawn apart ever since they were first resolved around 1780. Both are red dwarfs, less massive than the sun, and under a tenth its luminosity. Their spacing is approximately 80 AU. We know about a hundred thousand such multiple systems, mostly faint nearby stars. This is a 1980 photo by Tom Reiland and John Stein at the Allegheny Observatory, Pittsburgh, Pennsylvania.

10^{16} meters

1 Comets must be of the solar system, for they fall in to the sun with velocities short of entry speed from interstellar space. But they must be very loosely held to the sun, for they verge on the speed of escape. These speeds and the need for a large pool of the perishable items led to the postulate of the distant cloud, first suggested by Ernst Öpik and then worked out by Jan Oort.

From time to time some small gravitational impetus from a passing star starts a comet plunging down, down to the distant sun and back, a million-year voyage. That now happens a few times a year; so there must be 10^{10} or 10^{11} comets in the cloud, if they have been coming in during the whole of earth history. Most comets we see share that cold past, making only a single venture in to and back out of the warmth near the sun. The comets we catch sight of over and over again on earth are those few that have been trapped, mostly by unlucky gravitational encounters with Jupiter. These periodic comets gradually waste their substance through the growth and renewals of their tails, unless they are set free by another chance encounter, to escape again into the distant cloud.

2 The composite photograph of Barnard's Star is an enlargement from plates taken at the long focal-length 30-inch Thaw refractor of the Allegheny Observatory, in Pittsburgh, Pennsylvania, an instrument intended for this sort of work ever since it was built in 1912. Notice that only two stars are seen on the plate, which is exposed just enough to bring out the target star well. The comparison star is about ten minutes of arc away; it is a ninth-magnitude star in Ophiuchus. This whole visual presentation is something of a tour-de-force; astrometry is a technique not often taken out from behind a screen of numbers. We are grateful to George Gatewood, Director, and to Tom Reiland, who made the photographic study.

10^{15} meters

1 This illustration of Cassiopeia on her throne comes from one of the many editions of Hyginus, *Poeticon Astronomicon*, Cologne, 1534. Collection of Owen Gingerich.

2 The two bears are from an edition of Peter Apian, *Cosmographia*, Antwerp, 1540. Collection of Owen Gingerich.

3 The Chinese map of Orion is reproduced more fully and described in detail in the relevant volume of the living classic *Science and Civilization in China*, by Joseph Needham, with the collaboration of Wang Ling, Volume 3, Cambridge University Press, 1959. It is a version of a map prepared five hundred years earlier, itself based on older Chinese catalogues from the fourth century B.C. The drawing is taken from a manuscript found in Tunhuang by Sir Aurel Stein. Photo from the British Library.

4 The distances plotted for the main stars of Orion are found in *The Observer's Handbook 1981* of the Royal Astronomical Society of Canada, Toronto.

5 The Pleiades tangled in dusts not of their own making are from a Schmidt camera plate of the Harvard College Observatory, made at the Agassiz Station in Massachusetts, 1946.

10^{14} meters

1 When Halley's comet was last nearby, this photo was taken at the Boyden Station, then near Arequipa, Peru. It is a half-hour exposure on the blue-sensitive plates of the time (21 April 1910). Photo from the Harvard College Observatory.

2 Thomas Wright of Durham published in 1750 a theory of the cosmos in which, not without ambiguity, he interpreted the Milky Way as one disk of stars among others more remote. In the book he included "a true Delineation of the Solar System, with the Trajectories of three of the principal comets. . . ." The Comet of 1682, with a period of 75½ years, came duly around in 1758 after his book was out. Because that return had been predicted by Edmund Halley, first to recognize the periodic nature of some comets, the object has been called by his name ever since. The engraving is from *An Original Theory or New Hypothesis of the Universe*, by Thomas Wright, 1750.

3 The famous embroidery features the omen of Halley's comet. Courtesy of the Mayor of Bayeux.

4 The sky path of Halley was plotted by Roger Sinnott from the orbital formula published by Joseph Brady and Edna Carpenter, at the University of California, who took account of the effects of Jupiter. We have redrawn it from *Sky and Telescope*, February 1981.

5 The solar surface on 12 August 1917 shown in detail by the light of the strong red line of atomic hydrogen called Hα. Mount Wilson Observatory photograph.

Four spacecraft, two Pioneers and two Voyagers, have gotten so strong a boost from their encounters with Jupiter that now they are headed out of the solar system, to become the first material artifacts from earth ever to venture among the stars. Pioneer 10 will silently cross the upper edge of the small central square of 10^{13} meters during the late eighties, bound for the stars toward Taurus, though it will not leave the entire 10^{14}–meter frame for a century. Pioneer 11 is on its way out in almost the opposite direction with a few years delay; both Voyagers will be outside Pluto's orbit by 1990. All four carry explicit messages, like bottles tossed rather whimsically into the hyperocean of space.

Meanwhile, the edge of our possible influence moves out at full light speed. The farthest signs of our presence are the unintended result of powerful radar pulses and TV transmissions, enclosing a fast-expanding sphere but still well within the square of edge 10^{18} meters. See *Cosmos*, by Carl Sagan (listed in our Resources).

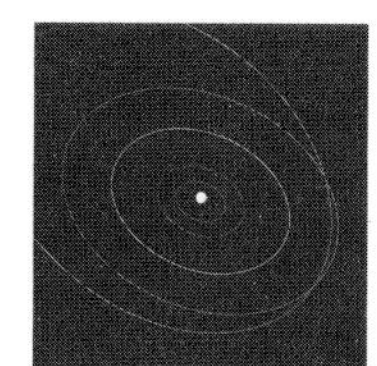

10^{13} meters

The National Aeronautics and Space Administration supports several centers whose photographic resources and expert aid have been generously given. They include the Jet Propulsion Laboratory (JPL), Pasadena, Ca., for planetary images; the Goddard Space Flight Center (GSFC), Greenbelt, Md., mostly for earth images; The Regional Planetary Data Center, Brown University Library, Providence, R.I.; NASA Headquarters, Washington, D.C., for images in general. We are grateful in particular to Leslie Pieri at JPL, Charles Bohn at GSFC, and Jim Head at Brown.

1 Voyager 1 took the dramatic picture looking back at Saturn on 16 November 1980. Voyager 1 will be tracked as long as possible; its power source is not solar but radioactive heat, so it has a chance to keep in touch. The photo is from NASA headquarters: number 80-HC-670.

2 The startling picture shows the complexity of Saturn's rings. Once they needed only simple names like A, B, C; Voyager 1 showed structure enough for a thousand or two distinct ringlets. Voyager 2 in 1981 seems to have shown another tenfold increase over that. No one doubted that the rings are made of a myriad circling bodies; but the presence of so many gaps and crowdings in the population

was not expected. Collisions, drifting dust, and lightning are here too. The photo was taken 6 November 1980, 8 million kilometers away from Saturn, by Voyager 1. A Jet Propulsion Laboratory picture: P-23068.

3 Titan is the major moon of Saturn; the other sixteen-plus, together with the rings, amount in all to less than one percent of Titan's mass. Titan is bigger than our moon. It is colder out there, and Titan can easily retain an atmosphere. It is dusty, as the scene shows, and perhaps it is loaded with colorful organic snows; nitrogen and methane are constituents of the atmosphere. Methane seas? The hemispheres differ in color, the north darker at this time. There must be some sort of global weather. The photo was made by Voyager 1 on 10 November 1980, 4.5 million kilometers away from Titan. It is JPL photo P-23076.

4 The ancient cratered ice sphere of Mimas is pictured from a distance of about half a million kilometers, on 17 November 1980. This is another JPL photo: P-23210.

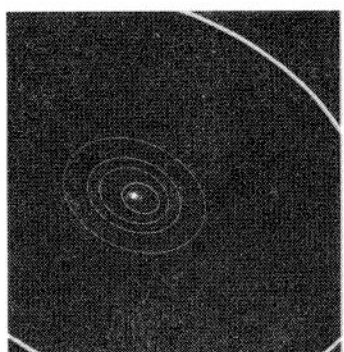

10^{12} meters

1 The Mt. Wilson 60-inch looks
2 at Jupiter. The fine detail is lim-
3 ited, mostly by air turbulence close to the telescope, to features on Jupiter about two thousand kilometers in size. But then send out Voyager 1 on a visit, bearing two TV cameras, one for wide and one for narrow view through unusually designed telephoto optics of no great size. Each TV camera is a sensitive vidicon, arranged for slow-scan readout, up to 48 seconds for each picture. They shoot through filter wheels that allow a choice of half a dozen color bands. Looking through the orange, blue, and green filters in quick succession gave the computers on earth enough information to build sharp color pictures. Voyager 2 brought the camera up much closer still, and we can admire the intimate result: the still-puzzling Great Red Spot. The whole planet from earth, Mount Wilson Observatory, ©California Institute of Technology; from space, JPL photo P-20993, Voyager 1, February 1979; the Red Spot, Voyager 2, July 1979, JPL P-21499.

4 Io was known to be out of the ordinary: Its orbit was seen both by Pioneer and from earth to be filled with gas. But no one expected the pyrotechnics of eight or ten ongoing volcanic eruptions as the Voyagers passed by. The energy probably comes from the varying tidal forces of massive nearby Jupiter, which knead and flex the moon, heating it throughout. (The radioactive heat that supplies earth's volcanoes is not enough there on little Io.) The plumes eject matter into space, to fill the orbital volume with sulfur ions, their glow visible from earth. Sulfur dioxide and sulfur must comprise the volcanic dust and lava for Io. (The grid of dots is a calibrating feature within the camera.) This is a JPL photo, P-21773, Voyager 2, 9 July 1979.

5 Galileo's notebook drawing of the dance of Jupiter's moons is from the Biblioteca Nazionale Centrale in Florence.

10^{11} meters

1 A Viking shot of Mars during approach in the summer of 1976. JPL photo P-19009.

2 A matched pair on Mars as seen
3 from Viking Lander 2, its camera nodding in slow mechanical scan of the scene. The place is the rocky plain called Utopia Planitia, north temperate Mars. The pictures were made on Mars days 871 and 960 after landing (15 February and 10 May 1979). JPL photos.

4 Venus at a distance of 59,000 km taken in February 1979 by Pioneer Venus Orbiter, during its first half a year in Venus orbit. This ultraviolet shot shows poorly understood markings high in that opaque atmosphere, probably swirling patterns of the sulfuric acid clouds. The high atmosphere of Venus spins many times faster than does the solid planet. This NASA HQ photo is numbered 79-HC-221.

5 Seven Soviet and four US probes have made their way to the forbidding surface of Venus. All surface photos have come from the Soviet landers; their first views go back to 1975, and the two latest probes gave us color photos in March 1982. The remarkable photo was taken on a rocky plain in the equatorial highlands of Venus. The surface sunlight looks orange under the thick clouds; some renewing geological processes are at work, for the landscape we glimpse is not worn smooth in spite of the

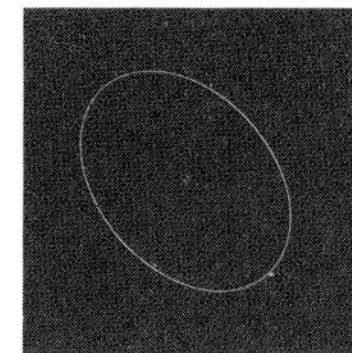

strong erosion of high pressure, high winds, and high temperature. Some fine-grained material is visible between the basalt rocks. That sawtooth ring is part of the lander, as is the color standard. Photo from Venera 13, USSR Academy of Sciences. Louis Friedman of The Planetary Society, Pasadena, helped to provide the photo.

6 Meteorite from the authors' collection.

The verse from Sumer is from the Iddin-Dagan tablets that bear sacred marriage hymns. The tablets with the verse seem clear evidence that by 1900 B.C. the identity of Venus as both Morning and Evening Star was clear in Sumer. The recognition of that continuity, a threshold for rational astronomy, antedates by a great deal the Homeric texts of the *Odyssey*, in which classical scholars claim to detect Greek ignorance on the point. This new translation of the hymn is the work of Diane Wolkstein and Samuel Noah Kramer, *Inanna, Queen of Heaven and Earth, Goddess of Love*, Harper and Row, New York 1982.

10^{10} meters

1 This is the first photo to catch crescent earth and crescent moon in one frame from so far away that the two appear in proper size relation. Voyager 1 took the picture on its way on 17 September 1977. The lunar image was brightened up threefold to make the moon easily visible on the print; the moon's gray dust is no match for the earth's reflecting white clouds. The photo is JPL's P-1981C.

2 This manuscript of Copernicus' book written in his own hand is held in the library of the Jagiellonian University in Cracow, where he was a student. The drawing occurs in Chapter 10 of Book I, early in the volume. Note that the circles are not scaled paths of the planets; rather, the planets were held by Copernicus to move within the spaces between adjoining circles. The photo is by Charles Eames.

Translation of Galileo's letter is by Stillman Drake.

10^{9} meters

1 Earthrise: taken from Apollo 17 in December 1972. NASA HQ photo 72-HC-976.

2 Moonrise: taken from Gemini 7 on 18 December 1965. NASA HQ photo 65-HC-618.

3 "*Having a pretty large corner of the Plate . . . void,*" Robert Hooke reproduced the moon crater which he drew "*by a thirty-foot Glass, in October 1664.*" The crater is near the center of the face of the moon; it had already been named Hipparchus by Riccioli, and the name has stuck. Hooke experimented with models for the origin of moon craters, but he decided against impact. Hooke, *Micrographia*. London, 1665. Photograph by Owen Gingerich. Permission of the Houghton Library, Harvard University. The Galileo remark is from his *Starry Messenger*, 1610.

4
5 The modern moon photo has been printed slightly out of focus for comparison with Galileo's original drawing, now held in Florence. "*His drawings are at least as good as one might expect from a non-artistic person (which he admits) using an unmounted, 20-power, poor-quality, narrow-field, dim-imaged telescope in cold midwinter with a flickering candle for illumination.*" Nevertheless Galileo's faithful hand was deft enough to allow after three centuries the identification of the exact time his drawing was made, "*about 5 p.m. Padua time on 2 December 1609.*" (See *Science*, Vol. 210. p. 136, 1980.) The comparison and the comments are due to Ewen Whitaker, Lunar and Planetary Laboratory, University of Arizona, whom we thank for the photos.

6 The snapshot shows the junior Senator from New Mexico, geologist Harrison H. Schmitt (R), on one of his more ambitious field trips, December 1972. The photographer was astronaut Eugene Cernan, and overhead in orbit astronaut Ronald Evans remained with their main vehicle. NASA HQ photo 72-HC-931.

7 The constellation Taurus can be seen, and perhaps also the stars of Corona. The photo is of an impression of a cylinder seal in the drilled style from late Assyrian times, carved between 1000 and 600 B.C. Boston Museum of Fine Arts, 65.1662.

10^8 meters

In 1948 Fred Hoyle looked ahead: "*Once a photograph of the Earth, taken from outside, is available—once the sheer isolation of the earth becomes plain, a new idea as powerful as any in history will be let loose.*" It took twenty years to confirm Hoyle's prescience. That idea was loosed by NASA before the end of 1967. First it was Lunar Orbiter 1, circling the moon, which sent a grainy and exciting image of the half-earth seen rising above a long lunar horizon, in August 1966. There were more such black-and-white photos from the next Lunar Orbiters. By January 1967 ATS-1, the first applications technology satellite, was out there in the geosynchronous orbit. Its TV camera showed almost the whole earth in black and white, day driving night westward out of the hemisphere. Then ATS-3 carried a new color scan camera into geosynchronous position over Brazil. Launched on 5 November 1967, five days later it sent back the famous first color picture of the whole earth, the blue marble of home against the black void. That image remains the most used of all in the NASA treasury; it is here in *2*.

1 Apollo 10: May 1969. The western United States, the north pole, and much of the Pacific Ocean are visible. NASA HQ 69-HC-487.

2 ATS-3: 10 November 1967. The whole earth; South America and West Africa are visible: NASA HQ 67-HC-723.

3 Apollo 11: July 1969. A watery view, showing the earth's north polar cap. NASA HQ 71-HC-104.

4 Apollo 11: July 1969. A crescent earth, showing the rosy edge of sunrise. NASA HQ 71-HC-379.

5 Apollo-ASTP: 20 July 1975. NASA HQ 75-HC-578.

6 Meteosat, the European Space
7 Agency's geosynchronous satellite, made these two images on 8 April 1979. The Europeans have placed their geosynchronous weather satellite above the crossing of the equator and the zero of longitude. The photograph at 6 shows an infrared image: The tones represent temperatures; cold, high clouds appear white, warmer clouds gray, and hot deserts show as very dark areas. Contrast in the spectral band (6 μm), where water vapor absorbs infrared radiation, produces a water-vapor map of the earth 7, taken at the same time as the preceding picture. Courtesy European Space Agency, Paris.

8 Apollo 17: December 1972. Antarctica, Africa, and the sunny Arabian peninsula can be clearly made out in this view of our planet. NASA HQ 72-HC-928.

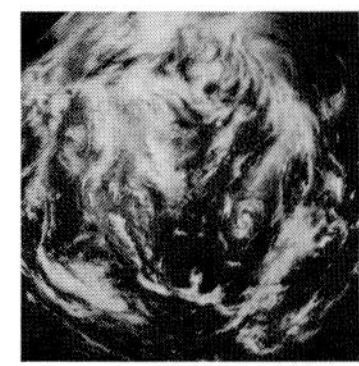

10^7 meters

1 This standard woodcut of the armillary sphere was long used to illustrate printed editions of that perennial bestseller, *De Sphera*, by John of Holywood (Sacrobosco). This reproduction is actually taken from a later rival work, *De Astrorum Scientia*, by Leopoldus, Augsburg, 1489. The publisher Ratdolt had simply reprinted a stock woodblock used for his own editions of Sacrobosco. Photo from the collection of Owen Gingerich.

2 The broken stratus is a familiar offshore cloud pattern. (The five-sided reflection is produced within the camera.) The picture was taken by an Apollo 11 astronaut on the way to the first moon landing in July 1969. NASA HQ photo 71-HC-114.

The diameter of a sphere is its quadrant arc multiplied by $4/\pi$, or by 1.27.

10^6 meters

1 The ice pack on Lake Michigan is driven downwind by the prevailing westerlies of winter storms. This Landsat photo was taken on 10 February in the cold winter of 1977. NASA HQ photo 80-HC-110.

2 The night photomap is routine nighttime coverage by the orbiting satellites of the Defense Meteorological Satellite Project. US Air Force photo, February 1974.

3 A hurricane of 2 September 1975 was caught by GEOS-1, the first geostationary one among the many successive meteorological imaging satellites. Photo from Goddard Space Flight Center.

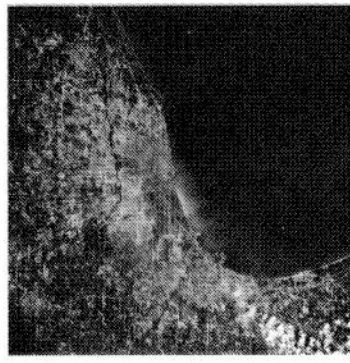

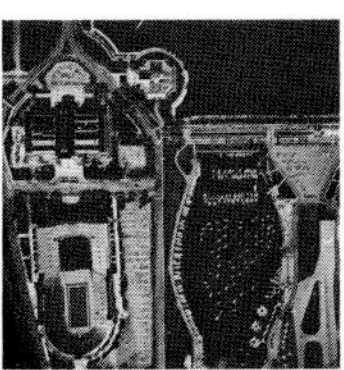

10^5 meters

A series of satellites, now generically called Landsat, began with a launch in July 1972. Landsat 3 is the one now in orbit; it passes over each place on earth about 9:30 in the morning. Each successive orbit shifts far westward; but next day the orbits run beside those of the day before, offset about 150 kilometers west. In 18 days the fields of view paint the whole earth from the 900-kilometer orbit height. Each photo includes a square on the ground of about 185 kilometers on edge. Detail is visible down to 80 meters. A powerful publication scheme has made year-by-year coverage widely available. The information is held on tape, but it can be prepared visually in many ways. The images are made in several color bands; their combination into color pictures is usual, though the bands available require a coded color presentation, the false color usual for infrared photography.

In general, Landsat images are available from Eros Data Center, Sioux Falls, South Dakota. Photos *1*, *2*, and *3* on this page courtesy Goddard Space Flight Center.

1 Here is a New Zealand winter scene as it was shot in August 1975, by Landsat 1. It has been reproduced in the handsome compilation *Mission to Earth*. (See "Resources.")

2 The Baluchistan desert in westernmost Pakistan along the Afghanistan border.The enhanced Landsat print is by courtesy of Ralph Bernstein, IBM, Palo Alto, California.

3 Hawaii, the big island, is present in a mosaic of two Landsat images made 11 February 1973, also found in *Mission to Earth*.

4 The Landsat photo of the Manicouagan crater in northern Quebec was made 20 April 1974. (See *Mission to Earth*.)

10^4 meters

1 Agricultural photography of Kansas fields by Grant Heilman, Lititz, Pennsylvania.

2 The fortress town of Palmanova, from J. Goldthorpe Aerofilms, Ltd., Boreham Wood, England.

3 Aerial photomosaic of Sioux Falls, South Dakota. Eros Data Center photograph.

4 Everest profile drawn from contours on the map in *The Times Atlas of the World*, Houghton Mifflin, Boston, 1967. Tonga Trench profile from *The Face of the Deep*, by Bruce Heezen and Charles Hollister, Oxford University Press, 1971.

10^3 meters

1 Pawnee star chart photo by Von Del Chamberlain, by permission of the Field Museum of Natural History, Chicago.

2 Orrery, by Benjamin Martin, photo by Roderick and Marjorie Webster, Curators of the Antique Instrument Collection, The Adler Planetarium, Chicago.

3 Soldier's Field Stadium, Chicago Park District photo, by Bud Bertog, December 1980.

4 The George Washington Bridge joins Manhattan at 179th Street with Fort Lee, New Jersey. It was designed by O. H. Ammann and opened in October 1931 by the Port Authority for whom Ammann was Chief Bridge Engineer. After fifty years, there are half-a-dozen longer bridges in the world; the longest of these is suspended across the Humber, about 1400 meters free span. But Ammann's daring bridge had doubled the span length of its predecessors. ©The Image Bank, New York.

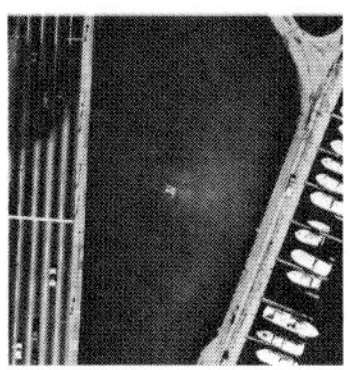

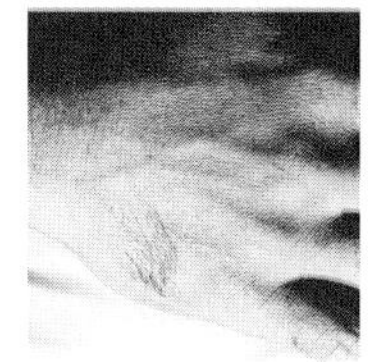

10^{2} meters

1 The General Grant Sequoia stands in Kings Canyon National Park, California. Photo in 1980 by David Muench, Santa Barbara, California.

2 An early test of the Saturn V, which is shown among gantries at Cape Canaveral. NASA HQ photo SAT 5-63.

3 The figure is based on a drawing on which the sculptor F. A. Bartholdi wrote "Praise to M. Eiffel with thanks from the Statue of Liberty who owes him her iron skeleton." Her exterior form is made of hammered copper sheets, rather heavier than the concealed frame. Redrawn from *The Tallest Tower*, by Joseph Harriss, Houghton Mifflin, Boston, 1975.

4 In Agra on the Yamuna in north India stands the serene tomb of the Empress Mumtaz Mahal built in her memory in 1632. The photo is from Editorial Photo Archives, New York.

5 The theater of Epidaurus is a "triumph of intelligible art" according to the archeologist who made the aerial photo with the aid of the Hellenic Air Force. See *Wings over Hellas*, by Raymond V. Schoder, S.J., Oxford University Press, 1974.

6 Barque-rigged, this graceful steel-hulled ship was launched at Hamburg in 1936; she became a war trophy of the US. Photo by US Coast Guard.

10^{1} meters

1 The elongated Jurassic giant, excavated near Vernal, Utah, by a Carnegie Museum team led by Carl Douglas in the 1920s, became one of the most celebrated of dinosaurs. Casts of the skeleton are found in museums around the world. The original stands in the Carnegie Museum in Pittsburgh, Pennsylvania (except that those last thin eight feet of tail are in storage). This beautiful reconstruction drawing is the recent work of Jim Senior of that Museum. Photo by the Carnegie Museum of Natural History.

2 In the course of building a lifesized reconstruction for the University of Nebraska State Museum in Lincoln, Roger Vandiver, now of Norman, Oklahoma, drew the lumbering rhino.

3 Photo of the printer's house is by Charles Eames.

4 Airplane, Cessna 152, Wichita, Kansas.

10^{0} meters

Protagoras is cited from the translation of the fragments in *Ancilla to the Pre-Socratic Philosophers*, by Kathleen Freeman, Harvard University Press, 1957.

1 A California sheep. Photo by Charles Eames.

2 A chair of plywood, model DCW, designed in 1946 by Charles Eames. Photo by Charles Eames.

3 Ripened wheat. Photo by Grant Heilman.

4 Topaz is a natural aluminum fluorosilicate. The outsized crystals came from Brazil. Photo 2054A by Smithsonian Institution, Washington, D.C.

5 The circus wagon wheel was photographed at the Circus World Museum, Baraboo, Wisconsin, by Charles Eames.

6 The photos of cheetah and pea-
7 cock are courtesy of the National Zoological Park, Smithsonian Institution, Washington, D.C.

10^{-1} meters

1 The portrait of St. Jerome is in the Church of All Saints, Florence. (The artist included the date in the painting.) Photo by Charles Eames.

2 Paul Revere's tools are at the Bostonian Society. Photo by Charles Eames.

3 Harrison built a succession of workable chronometers over thirty years; this one, his model Number 4, about five inches in diameter, is by far the most practical. It is kept at the National Maritime Museum, Greenwich, England; their photograph is here reproduced. On its trans-Atlantic test voyage in 1761, this chronometer kept time to a few seconds in a month, which implies an error in position of about one mile. "*I think I may make bold to say*," wrote Harrison, "*that there is neither any other Mechanism or Mathematical thing in the World that is more beautiful or curious in texture than this my watch. . . .*"

4 The evening primrose (*Oeno-*
5 *thera parviflora*) in two photos, one using a filter that passed only the ultraviolet and one without filter. Photographed by Thomas Eisner, Cornell University.

6 The fine frog (*Rana catesbeiana*) is by the edge of a pond in Rensselaerville, New York. Photo by Thomas Eisner.

7 The crustacean *Euphausia superba* Dana is the chief form of the Antarctic krill. Dense patches of the animals redden the surface waters there; other species of the genus are found in all seas. The century-old drawing by G. O. Sars, Norwegian marine zoologist, is from the *Report of the Scientific Results of the Exploring Voyage of HMS Challenger during the years 1873–1876*, 50 volumes, London, Edinburgh, and Dublin, 1885–1895. See Volume 13. Photo by Marine Biological Laboratory, Woods Hole, Massachusetts.

8 The shrew *Sorex longirostris*. Photo by John MacGregor from Peter Arnold, Inc., N.Y.

10^{-2} meters

1 An IBM typewriter keyboard for *katakana*. These signs derive from portions of Chinese characters; they use up all the standard keys and some shifted positions. Photo by Charles Eames.

2 The buttons are an inch across; they are held in the Adams National Historical Site in Quincy, Massachusetts. Photo by Charles Eames.

3 Mushroom on Martha's Vineyard, Massachusetts. Photo by Charles Eames.

4 The bombardier beetle, *Stenaptinus ignitus*, is just 2 centimeters long. That spray comes from a remarkable little internal explosion made by a sudden mixing of separately secreted oxidant and fuel within a special vessel of the beetle. Photo by Thomas Eisner and Daniel Aneshansley.

5 The marine organism is *Polyorchis haplus*, star of the 1970 film of the same name. Film and photo by Charles Eames.

6 This scanning electron microscope picture is made from a cellulose acetate cast replica of the skin of a fingertip. The apertures seen on the ridges are the openings of sweat gland ducts. The photo is from *Tissues and Organs*, by R. G. Kessel and R. H. Kardon, W. H. Freeman and Company, San Francisco. Copyright ©1979.

7 The soap foam was prepared and photographed by Cyril S. Smith, MIT, Cambridge, Massachusetts. The topological and mechanical analysis is found in a celebrated paper on such networks in *A Search for Structure: Selected Essays on Science, Art, and History*, by C. S. Smith, MIT Press, Cambridge, Ma., 1981. See also Peter Stevens, Chapter 7, cited in Resources.

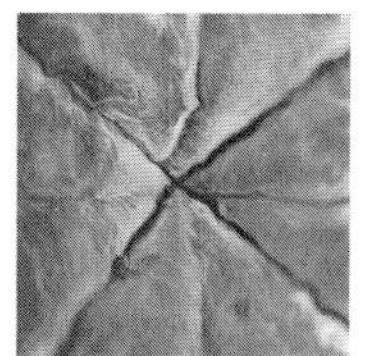

10^{-3} meters

The Hooke citation is found on the fourth page of his preface to *Micrographia*. (See Resources.)

1 The linen drawing is Hooke's,
2 found on Schem: XIII, Fig. 3, of the *Micrographia*. It is discussed in Obs. III. The poppyseed is Schem: XIX. Photos: Linen, by Owen Gingerich. Permission of Houghton Library, Harvard University. Poppyseed, by Charles Eames.

3 The big blue ciliate is seen alive in a phase-contrast optical microscope photo by Lynn Margulis, Boston University. The beads are part of the giant nucleus of this unusually large single-celled form, a protist, *Stentor ceruleus*.

4 These two are scanning electron
5 micrographs by David Scharf. The mite, order Acarina, probably an orabatid mite, is seen on a western subterranean termite. The salt was gold-coated for the exposure. See *Magnifications*, by David Scharf, Schocken Books, New York, 1978. Photo supplied by Peter Arnold Inc., N.Y.

6 The watch screw is shown in an SEM picture from the Eastman Kodak Company, Rochester, New York.

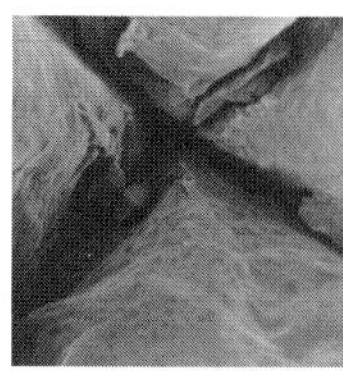

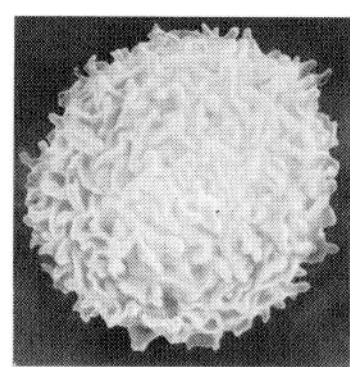

7 Richard P. Feynman, Professor of Physics, California Institute of Technology, offered a $1000 prize to "the first guy to make an operating electric motor that will be within a 1/64-inch cube, not counting the lead-in wires." His offer was made in December 1959 at a public meeting of the American Physical Society. William H. McLellan, an engineer in the laboratory of a Pasadena firm, then Electro-Optical Systems, designed and built a motor that met the conditions. It cost him a few months work on and off in the lab, with help from two technicians. The motor has four coils mounted on a square base. Each iron-core magnet coil is wound with the smallest commercial copper wire, 0.0005 inches in diameter. All other parts—disks, rods, and sleeves—were fabricated under the microscope (or two of them) with microdrill press and watchmaker's lathe. A sleeve bearing and a thrust bearing of quartz receive the molybdenum shaft with its magnetized disk. The motor is a two-phase synchronous AC motor; it works on milliamperes at millivolts, and has turned up to 1800 rpm. Data and microphoto from William McLellan of Pasadena.

8 This is the instruction decode unit of the two-chip general data processor of a microcomputer of the newest generation, from the Intel Corporation, Santa Clara, California.The chip (a high-performance metal-oxide semi-conductor) bears the layered logic circuitry of this firm. The nearly uniform black area is the sixty-four-thousand-bit read-only memory section. The light squares that border the chip are contact pads for its external connections. The chip is designated iAPX 43201. Microphoto courtesy Intel Corporation.

10^{-4} meters

1 This transverse section of the skin is an SEM photo from *Tissues and Organs*, by Kessel and Kardon.

2 Phonograph record grooves: SEM photo by Eastman Kodak Company.

3 Ernst Haeckel drew this elaborate radiolarian test in his copiously illustrated Volume 18 of the *Challenger* Report, Plate 22. It was taken in the Central Pacific, near the surface. Photo by Marine Biological Laboratory.

4 The name *radiolarian* is obsolescent, though still in common use. (These protists are now classified under the phylum Actinopoda.) The sample of tests has been acid-washed free of the sedimentary debris usual in the ooze of the sea bottom. This SEM is by John McNeill Sieburth, University of Rhode Island. It is found in his book *Microbial Seascapes*, University Park Press, Baltimore, Maryland, 1975.

5 The well-fed protozoan (*protist* is a current term) in the micrograph came from a wet ditch in Ithaca, New York. The filaments were once called blue-green algae; the present name for that group is cyanobacteria. The contoured relief effect is the result of the interference contrast optical system named after Nomarski. Photo by Thomas Eisner.

10^{-5} meters

1 The volume shown is within an arteriole, rather larger than the usual capillary. Blood runs about five million red cells—these little biconcave disks—to the cubic centimeter. Formed in the bone marrow, their life expectancy is a few months. They are cells specialized for transport that have lost their nuclei and most other organelles. The white cells are altogether more normal, though they, too, have a special immunochemical function. The deep SEM photo is from *Tissues and Organs*, by Kessel and Kardon.

2 The structure of the mammalian retina in cross section is unexpected. The network of signal-carrying nerve fibers, evidently transparent, covers the whole of the surface at which light enters. Beyond that layer the light continues through various cell connections and central bodies. Only at the far end of its passage is the light at last absorbed in the active layer of photosensitive dyestuff. There is a final backing layer of dark pigment, not preserved in this section. The SEM photo is by W. H. Miller, Ophthalmology and Visual Science, School of Medicine, Yale University.

3 The transmission electron micrograph (TEM) of the chloroplast is the work of Michael Walsh, Utah State University, Logan, Utah.

4 The wonderfully stacked pile of cannonballs is an SEM picture of a Gilson synthetic opal of high quality. The very regular spheres of silica have settled slowly but spontaneously into this configuration after a long time in water suspension. Photo courtesy of Pierre Gilson, Saint-Sulpice, Switzerland. (Compare with 10^{-9} m, Figure 3.)

5 The organism shown is a species of *Gonyaulax*, a close relative of the luminous principal marine organism of the toxic red tide. These plankton are dinoflagellates: They have two flagellae, whose placing results in a spinning motion through the water. The carved cellulose walls are distinctive for these protists. The SEM is by Eugene B. Small, University of Maryland. Courtesy Gregory Antipa.

6 This prototype circuit structure was made by a lithographic process in which the resist to be etched was exposed to a pattern made by x rays. SEM photo by Ralph Feder and Eberhard Spiller, IBM Research Center, Yorktown Heights, New York.

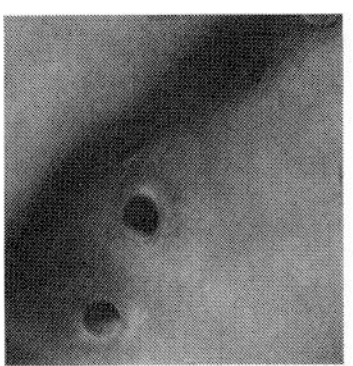

10^{-6} meters

1 The SEM picture of this freeze-fractured surface of the nucleus of a cell of onion is the work of Daniel Branton, Harvard University.

2 The TEM of a human chromosome in mitosis at late prophase shows the two complex DNA and protein strands of the newly duplicated chromosome joined at their crossing point. Micrograph by W. Engler, reproduced through the courtesy of G. F. Bahr, Armed Forces Institute of Pathology, Washington, D.C.

3 The TEM photo of this ubiquitous oxygen-breathing form, *Pseudomonas multivorans*, a weed among bacteria, magnifies a thin section. The cell wall is conspicuous; the dense grainy texture comes from the large number of protein-synthesizing ribosomes in the body of the swift-growing cell. In this sort of cell, the single unpackaged DNA loop is spread out on internal membranes. Photo by S. Holt and T. G. Lessie, University of Massachusetts. See *Five Kingdoms*, by L. Margulis and K. Schwartz, W. H. Freeman and Company, San Francisco, 1982.

4 The wavelengths of visible light arranged logarithmically. Under a hand lens the dots in any color halftone will suggest something of the additive nature of color perception. (See "Reading the Rainbow.")

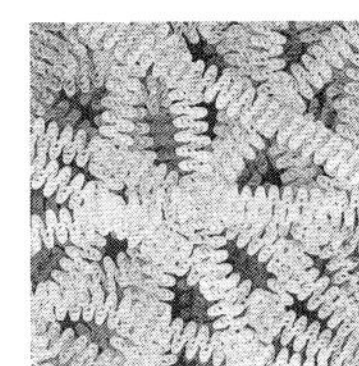

10^{-7} meters

1 The classic TEM photo shows a DNA molecule spilling from a broken virus casing, whose shape marks it as one of the best-known of bacterial viruses, T2 bacteriophage; that virus grows within the human colon bacterium, *Escherichia coli*. Photo by A. K. Kleinschmidt, Universität Ulm.

2 Here is a cluster of T2 bacteriophages attached to the hapless host bacterial cell. Some of the virus heads appear empty of their DNA content. The springs at the business end of several T2 particles can just be made out; they aid injection of the DNA molecular intruder into the cell. TEM by T. F. Anderson and Lee Simon, Rutgers University.

3 While an egg of the African clawed toad (*Xenopus* species) is being formed, there is multiplication of about 500 repeated genes, all of which encode for one particular group of RNA molecules. These molecules form key parts of the many ribosomes needed to carry out the very rapid protein synthesis that goes on to form the new egg cell. Some millions of transcribed RNA molecules are swiftly produced in the repetitive process so excitingly pictured here. The TEM was made by O. L. Miller, Jr., University of Virginia.

4 For this TEM the background substrate for the DNA-protein structure was stained with strongly scattering heavy molecules of uranyl acetate to increase the necklace visibility. The DNA string between the tiny beads of supporting protein has been stretched in the process. Photo by Ada L. Olins and Donald E. Olins, University of Tennessee, Oak Ridge Graduate School.

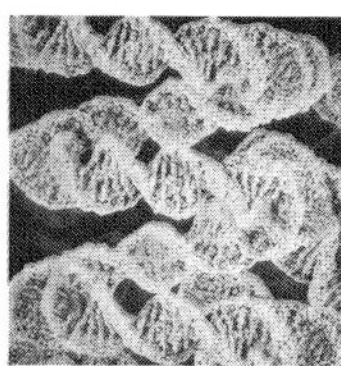

10^{-8} meters

The citation occurs at the end of Chapter 23 of *The Double Helix*, published by Atheneum in 1968 (paperback, Mentor, 1969).

1 Atoms visualized in a gas, a liquid, and a solid; the motion ubiquitous at this scale is called thermal motion. It provides the most profound measure of temperature. The molecules in all three of these samples of matter at room temperature move in a statistically well-defined way, with the same average speed for each, whether in free flight, in jostling exchange, or in tremorous vibration. The vibrating molecules within the crystal can never move far; they vibrate back and forth in a distance small compared to their own average spacing, so they must reverse direction more than 10^{13} times a second.

Thermal motion dominates the motions of the molecular world at ordinary temperatures. For larger objects it is sometimes detectable, though it is small; for smaller objects—electrons in the atom—it loses importance compared to the quantum motion.

2 The photo shows some molecules of the enzyme glutamine synthetase from the bacterium *E. coli*. An enzyme is a kind of molecular jig; its surface shape mediates and controls one particular step in a chemical reaction. Molecules that so mediate chemical reactions are called catalysts; *enzyme* is the term for a protein catalyst. Much of the text of the genetic message consists of recipes for enzymes.

This enzyme catalyzes the final step in assembly of the amino acid glutamine, which is an essential building block of many proteins. The enzyme participates in an elaborate feedback loop that fixes overall the rate of protein synthesis in the cell. The high-magnification TEM photo is by Earl Stadtman, National Institutes of Health, Bethesda, Maryland.

3 The TEM picture shows a transverse slice along one feeding microspine of a protist, a species of *Echinosphaerium*. The section shows about a hundred microtubules bundled into a neat spiral array to make the long spine. Each identical microtubule is itself a hollow tube of helically wound protein.

A photo of the whole handsome organism, a heliozoan or sun animalcule, similar to the radiolarians, is found in *Five Kingdoms*. The photo here is by J. A. Kitching, University of East Anglia, Norfolk, England.

10^{-9} meters

1 Dalton published these symmetrical molecules in his three-volume book, *A New System of Chemical Philosophy*, Manchester, 1808, 1810. There was no chemical evidence yet for the forms, and only partial evidence for the composition of the compounds. Photograph by Owen Gingerich. Permission of Houghton Library, Harvard University.

2 Molecular models in standard form are much used. An account of conventional space-filling models and other types is found in *Biochemistry*, by Lubert Stryer, W. H. Freeman and Company, San Francisco, 1981. Wonderful molecule drawings in space-filling style by the late Roger Hayward grace the earlier chemistry text *General Chemistry*, by Linus Pauling, W. H. Freeman and Company, 1954.

3 This high-resolution electron micrograph shows the block structure which is typical of this complicated metal oxide crystal. It includes a pattern of layers with deficient numbers of oxygen atoms; these deficient layers appear here in a kind of edgewise view as dark boundaries across the entirely regular atomic lattice. The photo was made in 1973 by S. Iijima, Arizona State University, Tempe.

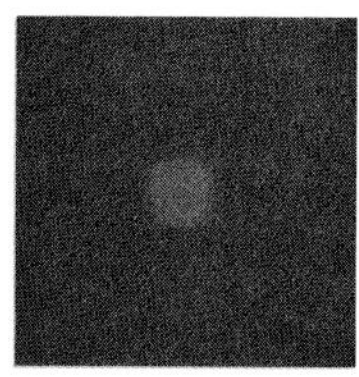

10^{-10} meters

1 This excerpt from the periodic
2 table of the elements and the
ionization data are information
from any general chemistry text.

The citation is by now part of Trinity College oral tradition; in a politer form it was published in *Rutherford and the Nature of the Atom*, by E. N. Andrade, Doubleday/Anchor, Garden City, N.Y., 1964.

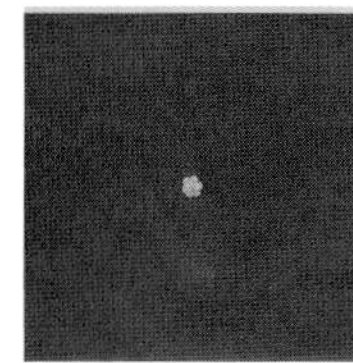

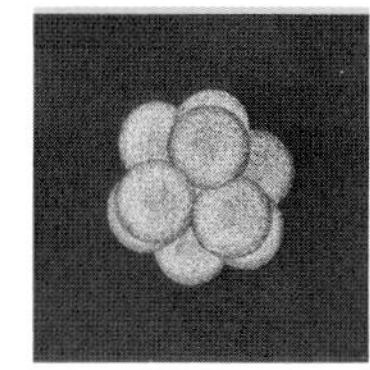

10^{-11} meters

1 The x-ray photo of a hand was made by Michael Pupin, renowned Columbia University professor of physics, as part of the first flood of x-ray interest. The patient was a New York attorney, his name now lost. Photo from the book *Röntgen and the Discovery of X-Rays*, by Bern Dibner, Franklin Watts, Inc., New York, 1968.

2 The graph of x-ray spectra is a form of the diagram first published by Henry Moseley in 1914. Moseley was killed in action at Gallipoli in the Dardanelles at the age of twenty-seven. See Segrè in the Resources.

10^{-12} meters

1 From this point down to smaller scales most of the evidence is so indirect and abstract that it is presented in diagrams and charts, rather than photographs. Bohr early worked out this physical account of the periodic table of the elements with his old quantum theory of electron states, before even the exclusion principle of Pauli was clear, not to speak of modern quantum mechanics. The table—empirically based—remains essentially correct. Note that the transuranic elements were foreseen as a set of chemically similar atoms. From *Annalen der Physik*, volume 71, 1923.

10^{-13} meters

1 The data for the isotope charts
2 of the elements are taken from *Tables of Physical and Chemical Constants*, 14th edition, by G. W. C. Kaye and T. H. Laby, Longman, London, 1973.

10^{-14} meters

1 The pulsing optical source in the center of the Crab Nebula was long known as a star with a strange spectrum; in early 1971 it was found to flash on and off. Soon after that this cinemalike view was made. It is no ordinary movie; the flashes are too faint to record without electronic tricks. The frames shown are spaced by about one three-hundredth of a second. Photo from Kitt Peak National Observatory.

2 The electron-positron pairs and
3 the 90-degree proton-proton collision are shown as parts of more complex events, in photos of a bubble chamber of liquid hydrogen exposed to a secondary beam at the Brookhaven National Laboratory proton accelerator in Long Island. In both cases the incoming particles are unstable positive pi-mesons with an energy of 3.9 GeV. Such mesons have relativistic speeds; they are moving at 99.93 percent of the ultimate speed; their mass during that motion is 27-fold what they have at rest. At this point relativity cannot be glossed over. Here it is fundamental. The big accelerators are examples of relativistic engineering on large scale. These photos (and a similar one at 10^{-15} meters as well) come from the extensive files of Robert Hulsizer and his group at MIT, one of many groups worldwide who measure and interpret large numbers of bubble-chamber photos.

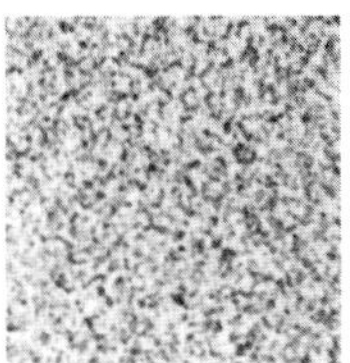

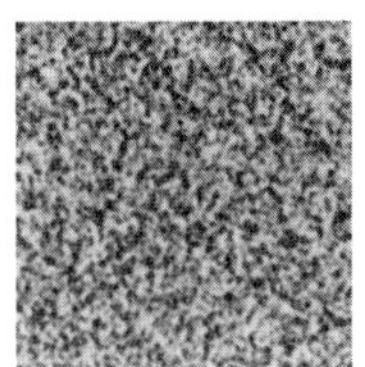

10^{-15} meters

1 This complex event is another bubble-chamber picture. Now, however, the bubble chamber with its liquid hydrogen is the one at the Fermilab, near Chicago (shown in the section Instrumental Vision). Here the incoming beam is at even higher energy; it is a mixed beam of protons with π^+ and K^+ mesons at 200 GeV. The electron has its antiparticle, the positron, and so also do all the other leptons and quarks. The antiparticles are so intimately related that they are not often regarded as types distinct from their particle counterparts. We tabulate the particle families as they are believed to be.

1. Quarks. Three pairs of quarks build up all the other strongly interacting particles, collectively called hadrons. Each of the six quark types or "flavors" (and its anti-quark) comes in three "colors," the name for a new analogue to electric charge.

2. Leptons. Three pairs of weakly interacting electronlike particles are called leptons. These include the stable electron and two more massive unstable relatives, each with its own neutrino type. The three lepton pairs are associated with the three pairs of quarks.

3. The field-mediating particles. These are exchanged among any particles in interaction; each kind of force requires its own specific carriers. The electromagnetic and neutrino-related forces are unified through the exchange of four related particles: the familiar photon or light quantum, and three heavy particles, not yet verified.

The strong interactions between quarks require the exchange of eight entirely novel particles called gluons, which carry color, and one or more neutral particles. (These are all indicated, but hardly verified.)

Gravitation requires a mediating particle, still undemonstrated.

10^{-16} meters

The Anchoress Julian of Norwich wrote this around 1400. The version here comes from *Revelations of Divine Love*, by Julian of Norwich, edited from a manuscript in the British Museum by Grace Warrick, Methuen & Co., London, 1901.

The formal qualities of the small-scale world of the particles echo the early uniformities of the great cosmos. Today we see hints of a direct physical relationship between the transient particles and the embryonic universe, once so hot that the novel particles, and not our familiar long-lasting matter, were its main contents.

THE IMAGES FROM THE FILM

Powers of Ten, A Film Dealing with the Relative Size of Things in the Universe and the Effect of Adding Another Zero, made by the Office of Charles and Ray Eames, is a color and sound film 9½ minutes long. The forty-two large square images that mark the powers of ten were rephotographed especially for reproduction in this book from the original images used in the production of the film. The work was done by Alex Funke, who had taken a major part in the 1977 production. He has prepared this brief technical note on the sources and techniques used in the film itself:

"*Powers of Ten* is essentially one continuous animation-stand truck shot. The animation was photographed in a series of ten-second moves, made in such a way that the apparent acceleration is constant. Every ten-second period begins with a big close-up of the center of a large image, and ends on a field ten times larger. At the center of each image—there were more than a hundred of them—was an inserted, ten times reduced view of the entire preceding scene, to assure continuity of detail and color. The successive moves were linked by in-camera dissolves.

Some of the pictures in this book show signs of their use in the film. In some cases the inserted center picture is discernable; and in almost every case, the edge of the square image will have less detail than the center. In order to retain the proper feeling of perspective, the degree of detail in each area of the image was made to be inversely proportional to the distance from which the area would be seen by the camera.

In preparing for the film, we first sought out at every power the very best pictures available, then asked workers in that particular realm what we might see if the imaging were a hundred, a thousand times better. We had the raw material—the aerial photographs and the shots from the Hasselblads of Skylab, the radio maps of hydrogen in the arms of our galaxy, the plates from the great telescopes, elegantly freeze-cleaved sections of leucocytes, and the vast mathematical models of large and small things, local groups of galaxies and clouds of electrons. Then in each case we made the imaging more than real through adding, by hand, the details of what might (or should) be there.

When there were only mental models, we made physical ones; built the doubly twisted DNA helix, animated the electron shells and the quarks. And we drew upon the mapmakers, the aerial civic surveyors, the dermatologists with their casts of the skin in silicone and in epoxy, the makers of the Lick Sky Survey, the scanning electron microscopists in half a dozen disciplines. We tried to pack into each final image enough information to give the illusion of almost unlimited layers of texture and detail."

A. F.

Here are the title and credits of the film:

Powers
of Ten

A Film Dealing with
the Relative Size of Things
in the Universe
and the Effect
of Adding Another Zero

Made by the Office of
Charles & Ray Eames
for IBM

©Charles & Ray Eames 1977

Music Composed
and Performed by
Elmer Bernstein

Narrated by
Philip Morrison

For the Eames Office:
Alex Funke
Michael Wiener
Ron Rozzelle
Dennis Carmichael
Wendy Vanguard
Cy Didjurgis
Don Amundson
Michael Russell
Sam Passalacqua

Consultation:
John Fessler
Owen Gingerich
Kenneth Johnson
Jean Paul Revel
and
Philip Morrison

With Thanks to
Chicago Aerial Survey
Graphic Films
Modern Film Effects
NASA
Norman Hodgkin

With Much
Gratitude to
Kees Boeke

A first version of the film was made in 1968. The title and credits are:
A Rough
Sketch
for a
Proposed
Film
Dealing with
the Powers
of Ten
and the Relative
Size of Things
in the Universe

With Much
Gratitude to
Kees Boeke

Research
and Development:
Judith Bronowski

Effects:
Parke Meek

Production:
Antii Paatero
Tadas Zilius
Ted Organ

Music Composed
and Performed by
Elmer Bernstein

Made for the Commission
on College Physics
by the Office
of Charles Eames

UNITS OF LENGTH

1 The photo of the public meter is by Christian Gavel, courtesy Dorothy Erlandson, Chambre de Commerce et Industrie, Paris, 1980.

2 The photo of the yard is by Jehane Burns, 1973.

3 The 10-centimeter length—which we hope is handy for readers—is the metric system realized: not an image. The system was first put forward in 1790 by Talleyrand himself; the scientist-founders of the metric system were Borda, Lagrange, Laplace, Condorcet, Monge, and Lavoisier. The scheme was formally presented to the world from Paris by an international committee in June 1799.

The present standardized metric international system of units, with the somewhat trendy designation SI (for Système Internationale), was first officially approved by delegates from a few dozen nations at the Eleventh General Conference of Weights and Measures at Paris in 1960.

READING THE RAINBOW

The prisms are set up to recall an important experiment described by Newton in his *Opticks*. The photo was made at the Office of Charles and Ray Eames, by Alex Funke.

INSTRUMENTAL VISION

1 The plan of Tycho's observatory is a woodcut in his work *Astronomiae Instauratae Progymnasmata*, 1588. Photo by Charles Eames.

2 Two of Galileo's telescopes are preserved at the Museum of the History of Science in Florence. The longer one is of wood covered with paper; its objective is of 26 mm aperture, its magnification 14. The other is of wood covered with leather; its objective is stopped down to 16 mm, magnification 20. Photo courtesy of Maria Luisa Righini Bonelli, Museo di Storia della Scienza.

3 The drawing by John Herschel of his South African observatory was reproduced as the lithograph shown here, frontispiece to his report *Results of Astronomical Observations Made during the Years 1834/5/6/7/8 at the Cape of Good Hope*, by Sir John Frederick William Herschel, Smith, Elder and Company, London, 1847. (The entire expense of the expedition and of its report was borne by John Herschel himself.)

4 The establishment of the observatory on Mt. Wilson high above Los Angeles was the work of George Ellery Hale, beginning in 1905. At first, transport was by mule and burro; by 1908, the 60-inch reflector was on the mountain. This photo made around 1916 shows the improved transportation of the

day that helped in construction of the Hooker 100-inch telescope so well-used by Edwin Hubble. The Mack truck with the jaunty crew is heavily laden; a later shot shows it and its valuable load in the ditch, fortunately without harm. Photo by Owen Gingerich from the files of the Mt. Wilson Observatories.

5 The detailed cutaway drawing of the 200-inch in its dome was made in 1938 by Russell W. Porter. Photo from Palomar Observatory.

6 Arecibo Observatory's 1000-foot radio dish is inland and uphill from the city of Arecibo in Puerto Rico. The countryside is well-developed karst, limestone that dissolved away and then collapsed to form deep basins between knolls. Unlike the Theater of Epidaurus, this bowl is set in a rich green landscape; the porous limestone subsoil drains well, or the dish would stand in water. The three concrete pylons that suspend the dish on its cables like a great hammock are above 150 meters tall. This observatory is part of the National Astronomy and Ionosphere Center. Photo courtesy Cornell University/NSF.

7 Hooke's microscope is from the frontispiece of *Micrographia*, in a photo by Charles Eames.

8 The proton synchrotron is at the Fermi National Accelerator Laboratory in Batavia, Illinois. Aerial photo by Fermilab.

CHRONOLOGY

Many sources were used. We mention the sixteen-volume *Dictionary of Scientific Biography*, edited by C. Gillespie, Charles Scribner's Sons, New York, 1970–1980, and the *History of Technology*, 7 volumes, edited by C. Singer, E. Holmyard, A. Hall, and T. Williams, Oxford University Press, 1954–1978. (There is a useful small history by Charles Singer listed in Resources.) A valuable specialized book is *Electronic Inventions and Discoveries* (second edition), by G. A. Dummer, Pergamon Press, New York, 1978.

THE BOOK OVERALL

Every book today, in particular such a patchwork of text and image as this one, requires the joint work of many professionals. In addition to those whom we have mentioned before in a specific context, we want to thank these, whose work shows everywhere:

Alex Funke, who so elegantly made the forty-two full-page photos.

Ippy Patterson, who drew and re-drew the original illustrations made especially for the book.

Genise Schnitman, who made the index and more.

IBM, whose support made the film possible.

Malcolm Grear Designers, who gave the book its visual form; it was the work of Pat Appleton, Malcolm Grear, Bill Newkirk, and Rick Pace, with design consultation by Tina Beebe at the Eames Office.

Howard Boyer, Linda Chaput, Andrew Kudlacik, and Carol Verburg, all indispensable to the editorial work.

Laura Argento, Betsy Dilernia, Barbara Ferenstein, Charles A. Goehring, Gary Head, Bob Ishi, Linda Jupiter, Neil Patterson, Peter Renz, and Mary Winters of W. H. Freeman and Company.

Jehane Burns, Dennis Flanagan, Owen Gingerich, Jack and Nita Goldstein, and Lynn Margulis, friends and critics of the text and its logic.

Mary Dawson, Grant Heiken, Jackie Kloss, Richard T. Miller, Carol Palmer, Yvonne Pappenheim, Marianne von Randow, Conway Snyder, Gerald Soffen, Bob Staples, Judy White, and Robert R. Wilson, for diverse and generous assistance.

The young people who over the years explored Kees Boeke's cosmic view.

RESOURCES FOR THE READER

These are resources we wish to share with the reader. More specific references will be found in the "Sources and Notes" section.

THE FILMS

Powers of Ten. Made by the Office of Charles and Ray Eames for IBM. 1977. 9½ minutes.

A Rough Sketch for a Proposed Film Dealing with Powers of Ten. Made by the Office of Charles and Ray Eames. 1968. 8 minutes.

These films are available in 16 mm sound and color, and on video media, from Pyramid Films, P.O. Box 1048, Santa Monica, California.

BOOKS THAT TAKE AN OVERVIEW

Cosmic View: The Universe in Forty Jumps, by Kees Boeke. John Day, 1957. The origin of the tenfold journey; the book was made with and for children of junior-high-school age, but its appeal is wide.

One Two Three . . . Infinity, by George Gamow. Bantam Books, New York, 1979. A classic of witty exposition, completed in 1961, treating mathematics, macrocosm, and microcosm.

Knowledge and Wonder: The Natural World As Man Knows It, by Victor Weiskopf. MIT Press, Cambridge, Mass., 1979. With a philosophical flavor.

Cosmos, by Carl Sagan. Random House, New York, 1980. Personal, richly documented, up to date.

Timescale, by Nigel Calder. Viking Press, New York, 1982. A tenfolding voyage through time.

ASTRONOMY

The Astronomical Companion, by Guy Ottewell. Published in 1979 by Guy Ottewell, Department of Physics, Furman University, Greenville, South Carolina, 29613.

Astronomy: The Cosmic Journey, by William K. Hartmann. Wadsworth Publishing Co., Belmont, Ca., 1978.

Ottewell and Hartmann have made two admirable surveys of astronomy at the introductory level. The first is original, graphically brilliant, less a textbook than a hand-drawn visual tour led by a devoted guide. The second is an outstanding text, comprehensive and insightful.

The Red Limit, by Timothy Ferris. William Morrow & Co., New York, 1977. An informal introduction to the step at 10^{26} meters and larger, aimed at cosmological issues beyond our journey.

Discoveries and Opinions of Galileo, by Stillman Drake. Doubleday & Co., New York, 1957. Texts and commentary on Galileo and his work by a leading Galilean scholar.

THE SOLAR SYSTEM IN PARTICULAR

The Solar System, by John Wood. Prentice-Hall, Englewood Cliffs, N.J., 1979. A wide look, at the college level.

NASA has many excellent popular publications on the various probes and their results. Two of special interest are:

The Martian Landscape, by the Viking Lander Imaging Team. NASA SP-425, 1978.

Voyage to Jupiter, by David Morrison and Jane Sanz. NASA SP-439, 1980.

THE EARTH

Mission to Earth: Landsat Views the World, by Nicholas Short, Paul Lowman, Jr., and William Finch, Jr., NASA SP-360, 1976. A big volume with hundreds of Landsat views in color, set into context.

The International Atlas. Rand McNally & Co., Skokie, Ill., 1969. A useful work for map references.

Continents Adrift and Continents Aground, edited by J. Tuzo Wilson. Readings from *Scientific American*. W. H. Freeman and Company, San Francisco, 1976. An anthology of the new geology.

Geology Illustrated, by John S. Shelton. W. H. Freeman and Company, San Francisco, 1966. Admirable introduction to geology with aerial views to support the text.

Field Guide to Landforms in the United States, by John Shiner. Macmillan Co., New York, 1972. With the classic maps drawn by Erwin Raisz.

A Field Guide to the Atmosphere, by Vincent Schaeffer and John Day. Houghton Mifflin Co., Boston, 1981. A guide to wind and water, rainbow and snowflake. Richly illustrated.

CLOSE TO HUMAN SCALE

Structures, or Why Things Don't Fall Down, by J. E. Gordon. Plenum Publishing Corp., New York, 1978. A fascinating introduction to forces and materials in structures from blood vessels to cathedrals.

Patterns in Nature, by Peter S. Stevens. Little, Brown & Co., Boston, 1974. Thoughtfully illustrated, this book summarizes at an introductory level the natural constraints on form, with ordinary perception as its main source of data. There is a good chapter on scale effects.

Grand Design, by George Gerstner. Paddington Press, New York, 1976. The world of nature and man by an artist of the aerial camera.

Any of the inexpensive factual annuals. We have used *The Hammond Almanac of a Million Facts, Records, Forecasts*; editor in chief, Martin Bacheller. Hammond Inc., Maplewood, N.J., 1980.

LIFE

Biology, by Helena Curtis. Worth Publishers, New York, 1979. A comprehensive and very well illustrated introductory text. See also *Biology of Plants*, by Peter H. Raven, Ray F. Evert, and Helena Curtis. Worth Publishers, New York, 1981.

Molecules to Living Cells, edited by Philip Hanawalt. Readings from *Scientific American*. W. H. Freeman and Company, San Francisco, 1980. An anthology of molecular biology.

Life Story, by Virginia Lee Burton. Houghton Mifflin Co., Boston, 1962. A book on evolution at a single location over time; a sort of *Powers of Ten* in time, for younger children.

THE MICROSCOPE

Micrographia, by Robert Hooke, 1665. Reprinted by Dover Publications, New York, 1961. An inexpensive reduced facsimile of a masterpiece of popular science.

The Scanning Electron Microscope, by C. P. Gillmore. New York Graphic Society, Boston, 1972. The electronic images of our day, a little in the manner of Hooke.

CHEMISTRY AND ATOMIC PHYSICS

Chemistry, by Linus Pauling and Peter Pauling. W. H. Freeman and Company, San Francisco, 1975. A clear introduction to the subject.

General Chemistry, by Linus Pauling. W. H. Freeman and Company, San Francisco, 1953. Older, but with unmatched drawings by Roger Hayward.

QUANTUM AND PARTICLE PHYSICS

From X-Rays to Quarks, by Emilio Segrè. W. H. Freeman and Company, San Francisco, 1980. A fine history of modern physics by a participant, who includes a good deal of the context.

Particles and Fields, edited by William J. Kaufmann, III. Readings from *Scientific American*. W. H. Freeman and Company, San Francisco. 1980. An anthology of particle physics.

Later accounts of the growing new theory of basic particles and their relationships are given in *Scientific American* articles: G. 't Hooft, June 1980; H. Georgi, April 1981; S. Weinberg, June 1981.

The Nature of Matter: Wolfson College Lectures 1980, edited by J. H. Mulvey. The Clarendon Press, Oxford University Press, New York, 1981. Eight well-known physicists survey the particles and their symmetries for nonspecialist readers.

THE METRIC SYSTEM

The Metric System: A Critical Study of Its Principles and Practice, by M. Danloux-Doumesnils. Athlone Press, University of London, 1969. A delightfully clear and candid account, including the history.

HISTORY

The Cosmographical Glass: Renaissance Diagrams of the Universe, by S. K. Henninger. The Huntington Library, San Marino, Ca. 1977. The enterprise of making visual models of the world during that decisive period of European history.

A Short History of Scientific Ideas to 1900, by Charles Singer. Clarendon Press, Oxford, 1969. Concise and yet general.

TO FIND MORE BOOKS

The Next Whole Earth Catalogue: Access to Tools, edited by Stewart Brand. Random House, New York, 1980. A thick, annotated guide to interesting books and other valuable tools, many bearing directly on our topics. Opinionated, free of inhibitions.

INDEX

Powers of Ten: The Video

The award-winning film on which this book is based, *Powers of Ten: A Film Dealing with the Relative Size of Things in the Universe and the Effect of Adding Another Zero,* made by the office of Charles and Ray Eames, is available on video cassette. The 21-minute video (The Films of Charles and Ray Eames: Volume 1, Pyramid Film & Video) also includes the original version of *Powers of Ten,* produced in 1968, entitled "A Rough Sketch for a Proposed Film Dealing with the Powers of Ten and the Relative Size of Things in the Universe."

The video cassette (ISBN 0-7167-5029-5) may be ordered through W. H. Freeman and Company by calling 1-800-877-5351 or by writing to: Marketing Department, W. H. Freeman and Company, 41 Madison Avenue, New York, NY 10010. The price is $39.95 + $1.95 shipping and handling (NY, CA, and UT residents should add applicable sales tax).

Hardcover books in the Scientific American Library Series

SUN AND EARTH
by Herbert Friedman

ISLANDS
by H. William Menard

DRUGS AND THE BRAIN
by Solomon H. Snyder

EXTINCTION
by Steven M. Stanley

EYE, BRAIN, AND VISION
by David H. Hubel

SAND
by Raymond Siever

THE HONEY BEE
by James L. Gould and
Carol Grant Gould

ANIMAL NAVIGATION
by Talbot H. Waterman

SLEEP
by J. Allan Hobson

FROM QUARKS TO THE COSMOS
by Leon M. Lederman and
David N. Schramm

SEXUAL SELECTION
by James L. Gould and
Carol Grant Gould

THE NEW ARCHAEOLOGY AND THE ANCIENT MAYA
by Jeremy A. Sabloff

A JOURNEY INTO GRAVITY AND SPACETIME
by John Archibald Wheeler

SIGNALS
by John R. Pierce and A. Michael Noll

BEYOND THE THIRD DIMENSION
by Thomas F. Banchoff

DISCOVERING ENZYMES
by David Dressler and
Huntington Potter

THE SCIENCE OF WORDS
by George A. Miller

ATOMS, ELECTRONS, AND CHANGE
by P. W. Atkins

VIRUSES
by Arnold J. Levine

DIVERSITY AND THE TROPICAL RAINFOREST
by John Terborgh

STARS
by James B. Kaler

EXPLORING BIOMECHANICS
by R. McNeill Alexander

CHEMICAL COMMUNICATION
by William C. Agosta

GENES AND THE BIOLOGY OF CANCER
by Harold Varmus and
Robert A. Weinberg

SUPERCOMPUTING AND THE TRANSFORMATION OF SCIENCE
by William J. Kaufmann III and
Larry L. Smarr

MOLECULES AND MENTAL ILLNESS
by Samuel H. Barondes

EXPLORING PLANETARY WORLDS
by David Morrison

EARTHQUAKES AND GEOLOGICAL DISCOVERY
by Bruce A. Bolt

THE ORIGIN OF MODERN HUMANS
by Roger Lewin

THE EVOLVING COAST
by Richard A. Davis, Jr.

THE LIFE PROCESSES OF PLANTS
by Arthur W. Galston

IMAGES OF MIND
by Michael I. Posner and
Marcus E. Raichle

THE ANIMAL MIND
by James L. Gould and
Carol Grant Gould

MATHEMATICS: THE SCIENCE OF PATTERNS
by Keith Devlin

Scientific American Library books now available in paperback

POWERS OF TEN
by Philip and Phylis Morrison and the Office of Charles and Ray Eames

THE DISCOVERY OF SUBATOMIC PARTICLES
by Steven Weinberg

THE SCIENCE OF MUSICAL SOUND
by John R. Pierce

THE SECOND LAW
by P. W. Atkins

MOLECULES
by P. W. Atkins

THE NEW ARCHAEOLOGY AND THE ANCIENT MAYA
by Jeremy A. Sabloff